CARDIFF
CAERDYDD

D0279494
ACC. No: 03191212

Island on Fire

First published in 2014 by
Profile Books, 3A Exmouth House
Pine Street, Exmouth Market
London, EC1R 0JH
www.profilebooks.com

10 9 8 7 6 5 4 3 2 1

Printed and bound in the UK by Clays, Bungay, Suffolk.

Typeset in Minion to a design by Henry Iles.

A CIP catalogue record for this book is available from the British Library.
ISBN 978 178125 0044
eISBN 978 184765 8418

Island on Fire

The extraordinary story of Laki, the forgotten volcano that turned eighteenth-century Europe dark

Alexandra Witze & Jeff Kanipe

PROFILE BOOKS

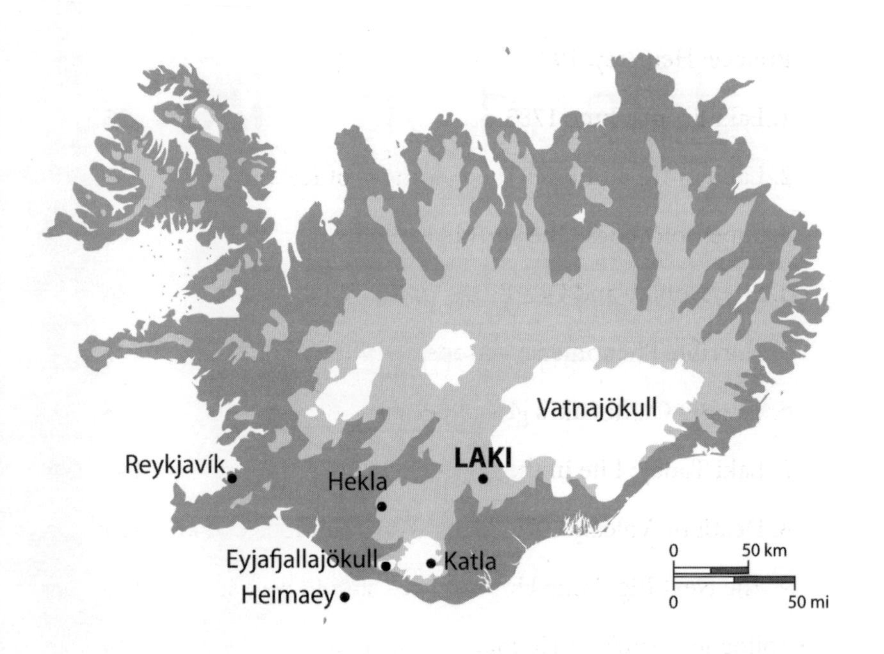
Vatnajökull
LAKI
Reykjavík
Hekla
Eyjafjallajökull
Katla
Heimaey
0
50 km
0
50 mi

Contents

PREFACE

Heimaey, 1973

AT LEAST EVERYONE WAS AT HOME, snug in their beds, when the world began to end.

It was late January, 1973, and the harsh Icelandic winter had been even rougher than usual on Heimaey, an island off the country's southern coast. On a normal night, Heimaey's fishermen would be out plying the rich waters of the North Atlantic for cod, haddock and herring. But on this night, high winds and stormy seas had kept the crews off their boats. Instead, they hunkered down in the trim little clapboard houses that sprawled above the island's spectacular black-cliffed harbour. So when the earth ripped open on 23 January, nearly all of Heimaey's 5,300 residents were there to see it.

It began just before 2 a.m., in a field on the island's eastern side. Only 200 metres from a serene little hamlet called Kirkjubær, or Church Farm, a line of flame spouted from the ground. It looked like a fire in dry grass, but of course it wasn't: it was the Earth's crust coming apart, spewing fountains of lava into the air. Within minutes the spouts of fire were 150 metres high. Within hours the rift was 1.5 kilometres long, almost splitting the island in two.

Had the eruption begun just a few hundred metres to the west, the volcano might have incinerated islanders as they slept. As it was, most of them found out about it when neighbours or police started pounding on their doors. Dazed, they staggered outside and stared at the fiery jets as close as the end of their street. Then they turned their backs, went inside and started packing.

In those first confused hours, no one knew what might happen to Heimaey. Would the eruption ignite the oil tanks down by the harbour? Would the entire town be engulfed by lava? Families grabbed what they could and made their way to the waterfront. One musician reportedly left wearing his pyjamas, clutching a frozen leg of lamb. Farmers at Church Farm shot their cows to spare their suffering, then fled.

Iceland's civil patrol sprang into action. They had practiced for just such an emergency, and the evacuation went swiftly and smoothly. Between 3 a.m. and 7 a.m. nearly all of the island's residents left Heimaey – via the fishing fleet that fortuitously had been left in the harbour, or via planes sent from Reykjavík. Those who left knew they might never return. Those who stayed knew they faced a battle like nothing before.

Seen in retrospect, it's not surprising that the eruption happened where and when it did. Heimaey is the biggest in a chain known as the Vestmannaeyjar or Westman Islands, named after the Celtic 'west men' who fled here in the ninth century after murdering one of Iceland's first permanent settlers. Volcanism is a way of life here; the islands are the tips of mostly dormant volcanoes poking above the waves.

But not entirely dormant. In November 1963, the Westman Islands increased by one, when new land rose above the sea southwest of Heimaey. This was Surtsey, built up from the ocean bottom by successive eruptions. The first people to spot the plume of smoke, from a nearby fishing vessel, thought it must be a ship on fire. But soon a fresh island appeared above the water, and eruptions built it up sporadically over the next three and a half years. Surtsey soon became a natural

Man vs lava: giant pumps sprayed seawater onto Heimaey's advancing lava flow.

laboratory for the study of how plants and animals colonize a barren, newborn terrain.

The Heimaey eruption, a decade after Surtsey, would come to occupy a special place in the annals of volcanology. After all, it's not every day that the Earth's crust is blown apart directly beneath a town, forcing the evacuation of nearly the entire population. More significantly, the islanders who stayed decided to fight back. The idea of battling a volcano may sound crazy, but humans have tried controlling the flow of lava before. In 1942, the U.S. military dropped bombs at strategic spots along a river of molten rock that was threatening Hilo, Hawaii, hoping to divert it away from the city. (The eruption stopped at around the time of the bombing, so although Hilo was saved the military couldn't take credit for it.) In Italy, over the years, authorities have also built earthwork barriers to try to divert Mount Etna's lava from the surrounding villages. But what happened on Heimaey dwarfed anything ever tried in Hawaii or Italy. Icelanders knew they would have to fight the volcano or abandon the island altogether. And that would have meant giving up the country's most productive, and most lucrative, fishing port.

To those who stayed on Heimaey after the evacuation, it became clear within days that something would have to be done. A huge cinder cone had already risen 150 metres above Church Farm. The volcano was spewing huge amounts of black ash and rock fragments, which buried half the town. Photographs from the time look like stills from a disaster film: men stand on rooftops, desperately shovelling the ash before it can accumulate enough weight to crush the buildings. Groups of people are silhouetted in front of burning houses and geysers of fire. Flames consume a hotel; black ash covers everything except the roof of a house. (Islanders saved dozens of homes by barricading windows and doorways with galvanized steel, to keep out flaming debris.)

Meanwhile, the half of Heimaey not smothered by ash was rapidly being covered by lava. More alarming still, the lava was heading directly for Heimaey's harbour. If it got there, it would cool and solidify into a huge natural dam, blocking the port. If that happened, Heimaey might as well shut down entirely, for without the fishing harbour the island was nothing.

That's when Thorbjörn Sigurgeirsson, an Icelandic physicist who had trained with Niels Bohr, came up with a plan for stopping the lava. His tools would be not bombs but water – seawater sprayed on the advancing flow to cool and slow its movement. Maybe, Thorbjörn said, workers could even redirect the lava to less crucial areas, away from the harbour.

Fifteen days after the eruption had started, the battle for Heimaey began in earnest. Bulldozers piled ash and dirt near the northwest margin of the lava flow, to create protective ramparts and divert it from town. Firemen positioned water pipes along the ramparts and sprayed seawater onto the creeping mass. And to almost everyone's surprise, the cooling plan seemed to work. The lava, chilled and thickened, went mostly where the firemen wanted it to go.

This struggle, however, could do nothing to staunch the flow at its source, and the eruption continued unabated. Some of the lava broke away and ran directly into the sea, dangerously close to the harbour entrance. Thorbjörn's team now enlisted two boats to pump more seawater onto the lava, but one of those boats, the *Sandey*, had been built for dredging sand, and its high-volume pipes were so large that a bulldozer would be required to position them across the flow. An intrepid driver duly did what was necessary. This might have been the first time anyone had ever driven heavy earth-moving equipment across flowing lava.

By mid-March, Heimaey looked like a battle zone: a network of pipes and hoses sprawled across the black mountainside, all the way down to the harbour. Wherever the water poured onto the lava it generated huge clouds of steam, so workers had to crawl atop a shifting mass of incandescent lava, feeling their way through a dense and hot fog. One team quickly got the nickname of 'the suicide squad'. (Amazingly, the only person who died in the eruption was not a lava worker but someone asphyxiated by volcanic gases in a poorly ventilated basement.)

Soon, by default, the Heimaey warriors became the world's experts on cooling lava. They learned that huge amounts of water were needed to stop the inexorable flow. The U.S. government sent more than 30 additional industrial-sized pumps, which let workers spray about 100 metres higher than before. In the meantime, Thorbjörn's team discovered that although plastic pipes would melt if empty, those filled with flowing water stayed cool enough to be laid intact across the lava surface. Workers hauled new pipes out onto the flow, where the plastic shifted safely with the flowing lava rather than fracturing as metal pipes had done.

Ship traffic today can make it through the narrow channel approaching Heimaey's harbour, thanks to the cooling and stopping of the 1973 lava flow (foreground).

At the peak of the lava cooling, some 75 men worked on the project. Most worked 13 days in a row, from 8 a.m. to 10 p.m., and then took two days off. Supervisors worked one week in Heimaey, then had one week off on the mainland. Occasionally, the spraying did more harm than good. Several times, a chunk of lava broke from the water-cooled dam and started drifting downstream. The biggest of these chunks, dubbed Flakkarinn or 'The Wanderer', weighed some 2 million tonnes. Thorbjörn's team figured out where Flakkarinn was heading and focused their spraying there. When the Wanderer smashed into the cooled lava wall, the errant mass broke apart.

By Easter week, lava had stopped flowing westward toward town, and on Easter day it moved safely eastward toward the sea – from then on, this was the only direction it ever went. By the time Heimaey's eruption was over, five months and five days after it began, the volcano warriors had pumped more than 6 million cubic metres of seawater, and the flow had stopped less than 100 metres from the quayside. Most of the lava was diverted to the north and east, where it spread out and added nearly 20 per cent to the island's area. Still, more than 400 homes were destroyed, and one-third of the town obliterated.

Drive across Heimaey today, and you can see how dramatically the island was reshaped by the 1973 eruption. Instead of one cinder cone dominating the landscape above town, there are two. The new one, Eldfell or 'Fire Mountain', has already oxidized to a rusty brown colour; soon grasses will cover it just like the older peak, Helgafell. Look down toward the harbour and you can see fresh black lava piled up against the back yards of houses. On the east side of town, residents have kept one reminder: a street buried in cinder ash, where you can peer into the remains of buildings destroyed during the eruption. Windows gape open, blown out by volcanic fire. Roofs crumple under the weight of black ash. Pompeii of the North, they call it.

Closer to the harbour, though, you can drive across roads freshly cut into the new lava and even visit a 'lava garden' with flowers, statues, pinwheels and other sun-bleached, wind-battered decorations. At the harbour itself you can stand on the edge of the 1973 lava flow and watch ships pass directly in front of you, through the narrow channel that still provisions Heimaey's harbour. On the cliffs opposite, puffins call to one another.

Heimaey has eked some good things out of the eruption. Ash was used to add a second runway at the island's airport and as landfill for new homes. The harbour is also easier for ships to manoeuvre into now, since it has a natural breakwater formed by a shallow underwater flow near its entrance. The high cone of Eldfell shelters the town from the strong winds that used to rake it. And for nearly a decade after the eruption, Heimaey residents plugged directly into the geothermal heat fuelling the volcano and used its energy to light their homes.

In the end, Heimaey represents the quintessential struggle between humans and nature, between engineering know-how and natural disaster. It shows Icelanders at their very best, rallying to save their homes and livelihoods from almost certain destruction. But our story is about another confrontation with lava, one with a very different outcome. It begins nearly two centuries earlier, to the north and east of the Westman Islands.

We begin in the heart of Iceland's volcanic fire, and with one extraordinary man who lived there.

CHAPTER ONE

Laki Erupts

June 1783

AROUND 9 A.M. ON SUNDAY, 8 JUNE 1783, Reverend Jón Steingrímsson stepped out of his small farmhouse, mounted his horse and began the five-kilometre journey to church. Sunday services were always his favourite part of the week, but he was particularly looking forward to today's observance. It was Pentecostal Sunday, also known as Whitsun, which commemorates the appearance of the Holy Spirit among the disciples of Christ after his resurrection. It was to be a day of celebration and reflection, and Jón was expecting his little Lutheran church to be brimming with worshippers.

For the past five years, as priest of the Sída district in southern Iceland, the 55-year-old clergyman had overseen what he considered to be a happy and prosperous spiritual flock. Jón and his family too had prospered, so much so that his farm was nearly overrun with sheep and cattle. His affluence had allowed him to host expensive weddings for two of his daughters, including a substantial dowry for each. He was the very model of a successful rural preacher.

Jón knew well how strenuous life in Iceland could be. But on this bright and clear Whitsun morning, with sunlight playing

Laki erupted in 1783 along a long straight fissure, similar to the fire fountains seen here in a 1984 eruption of Mauna Loa, Hawaii.

over the lush pastureland and the sheep and lambs grazing among the wildflowers, God seemed to be smiling. Then Jón chanced to look northward over his shoulder, and abruptly his reverie dissolved. He pulled up his horse and gazed in wonder and alarm. Looming over the foothills was an enormous, roiling black cloud.

This, Jón thought, was the end. The earthquakes that had shaken the ground over the last few days, some strong enough to frighten people out of their homes, had been but a clamorous prelude to something he had grimly foreseen. God's patience had run its course; the hour of afflictions had arrived. Whitsun would not be a day of celebration, but one of weeping and lamentation.

Within minutes the cloud was so thick it shut out the sun and drove everyone indoors, where even lamplight could barely dispel the enfolding darkness. People caught outdoors had to grope their way home in the blackness. Soon, a blizzard of powdery fluff began falling out of the cloud, settling thickly on the ground like coal ash. A light drizzle followed, turning the flakes into an inky slurry.

A brief respite came later in the afternoon, when a southeasterly sea breeze drove the ash cloud back across the foothills. Jón managed to conduct his services under a clear sky, but the relief was only momentary. That night the earthquakes returned. Then all hell broke loose.

In the days that followed came more tremors, more ash and cinder falls, darkness, filthy air and bitter acidic rains that burned the eyes and skin and scorched the pastures. Thick haze rolled across the countryside, accompanied by a devilish stink. Pastoral Iceland, once full of lush grassy meadows, became a grey and poisonous place.

At the time, neither Jón nor anyone else knew that the source of these earthly convulsions was the eruption of a new volcanic fissure system in the Icelandic highlands, some 45 kilometres to the northwest. Later the system would be given a name: Lakagígar, or the Laki craters, after Mount Laki that stands at the centre of the fiery seam. (Throughout this book, 'Laki' will be used as shorthand for both the Laki crater row and its 1783–84 eruption.) Laki erupted for eight months, and eventually, indirectly, it killed at least half of the country's livestock and one-fifth of its people. Icelandic historians would come to consider the eruption the single most devastating event since Víkings settled the island in 871 C.E.

The terrible irony was that the people in Jón's district had just clawed their way back to relative prosperity. A decades-long string of disasters had begun in 1750 with bitterly cold temperatures. Sea ice congealed around the coasts, preventing farmers from open-boat fishing during the winter, a practice that normally sustained them through the off season. Livestock perished, villages were devastated and several thousand people starved to death. As if that weren't enough, the volcano Katla erupted in the autumn of 1755, destroying much of the pastureland with ash fall and floodwaters. The weather improved slightly thereafter, but a smallpox outbreak ravaged the country in 1760, followed by scabies, which in eighteen years slashed the country's precious sheep population by nearly half.

Since then, however, the climate had moderated, the epidemics had diminished and Icelanders had enjoyed great bounties from the land. Good fishing had returned, the hay meadows were lush again and everyone had more than enough livestock for milk and meat. The country's population, which had dwindled to 43,000 in the depths of the famine, now surpassed 50,000.

But this surfeit, which Jón believed should have had people dropping to their knees in gratitude, seemed to make them more self-absorbed and uncharitable. Even as they prospered, he noted acidly, they became increasingly arrogant, lazy and debauched. Farmers had so many sheep that they gave up counting them. Food was so abundant that even servants turned up their noses at any but the richest victuals. The consumption of tobacco and alcohol increased: 'During a single year here,' Jón later wrote, 'spirits amounting to the worth of 4,000 fish were consumed at feasts, visits and the like.' Some clergymen showed up drunk to lead services. Their behaviour reminded Jón of a saying by his favourite Roman poet, Ovid: 'More often than not men fall into excess when their affairs run smoothly.'

To Jón's way of thinking, the impending disaster should have surprised no one. The supernatural portents were there for anyone to see. A fiery redness had appeared in the sky, and several years earlier 'monsters of various shapes' had been spotted in a stream to the south. 'Fireballs lay in heaps like foxfire' in a nearby coastal village, and 'noxious flying insects' moved across the land, he wrote in a chronicle describing the

Jón Steingrímsson's church in Klaustur would have looked much like this traditional building at the folk museum in Skógar, on Iceland's southern coast.

eruption. Wailing could be heard coming from the bowels of the Earth, and ethereal bells rang through the air. More than the usual number of lambs and calves had been born deformed. In one case, Jón reported, a lamb on a nearby farm was born with the claws of a predatory bird instead of hooves. Elsewhere, horses were said to feed from dung heaps. Time and again Jón had warned his parishioners that such omens preceded misfortune.

Just as God sent signs to people when they were awake, he also spoke to them when they slept, or so Jón believed. He put great stock in the predictive power of dreams, including his own. In the spring of 1783, many signalled that great changes were coming.

In one of his dreams, Jón wrote, a man appeared in a house full of farmers who were making merry and singing over their drinking cups. When they offered the stranger a cup and asked him to sing a 'merry ditty,' he cried out in anguish: 'The sun! The sun! The sun! Doomsday will soon be upon us!' The farmers rebuked the stranger for being so gloomy. But in his dream, Jón told the farmers not to mock him. Jón then asked the stranger his name, and he answered that he was called Eldrídagrímur. ('Eldur' is Icelandic for 'fire'.) Jón asked if he had visited the area before, and the man replied that he had, 'in the year 1112'. Jón's dream apparently ended there, but later, when he consulted a book of annals, he learned that in that year great lava flows had devastated the land (possibly a reference to the 1104 eruption of Hekla).

In another dream, a 'regal' figure approached Jón and advised him to preach the thirtieth chapter of Isaiah. At the time, Jón was feeling guilty for having to cancel church services nine Sundays in a row because of bad weather. But the visitor told him that the following Sunday would have fair weather – which, apparently, is how it turned out. Just in time, it seems, for Jón to convey his message. Chapter 30 of Isaiah opens with God's rebuke to the kingdoms of Jerusalem for believing that they no longer need him, and concludes with the Lord's terrible judgment of the Assyrians:

> *And the Lord shall cause his glorious voice to be heard, and shall show the lightning down of his arm, with the indignation of his anger, and with the flame of a devouring fire, with scattering, and tempest, and hailstones ... For Tophet is ordained of old; yea, for the king it is prepared; he hath made it deep and large: the pile thereof is fire and much wood; the breath of the Lord, like a stream of brimstone, doth kindle it.*

Jón's interpretations of his environment and his dreams were not purely theological – he had an abiding interest in the natural world as well. After all, he had been born and raised in one of the most volcanically active countries on the planet. Like many Icelanders, Jón had witnessed eruptions and their effects, such as the great glacial floods that rushed down from the ice-buried Katla in 1755. Unlike many Icelanders, however, he also noted how ash from these historic eruptions was distributed around the country, so that in one place there might be five ash layers in the ground and in others eleven or more. He saw these layers as being a kind of historical record, relaying stories of the severity of earlier eruptions and, by inference, the human costs. Today this is standard volcanological wisdom, but in the backwoods of Iceland in 1783 it was cutting-edge.

To be sure, it didn't take an observer as keen as Jón to see that some sort of volcanic event was in the offing. In the spring of 1783, a bluish smoke had been observed drifting over the ground, though no one knew its source. In early May, the crew of a Danish brig sailing around Iceland's southern coast claimed to have seen columns of fire shooting in the sky in the mountains north of Jón's district. Some weeks later, the Skaftá River rose to uncommonly high levels, breaching its banks violently in places, and the water was muddy and foul-smelling rather than clear and fresh. And then there were the violent earthquakes in the days leading up to Whitsun.

To Jón, the impending cataclysm would be known as the 'Fires,' the 'Scourge,' or even 'God's chastisement.' Within Iceland the eruption also goes by the name Skaftáreldar ('Skaftár fires'), because the main flow of lava more or less followed the gorge of the Skaftá River to the southern coast. Modern geologists refer to the disaster as Lakagígar (if they are Icelandic) or Laki (if they are not), and consider it to be one of the most significant eruptions of the modern era. It changed both the history of Europe and the history of science. Yet most people outside of Iceland have never heard of it.

Why does Laki remain so obscure? One explanation is that it does not fit the modern concept of what a volcano should be: unlike Italy's Mount Vesuvius or Japan's Mount Fuji, it does not tower scenically above a well-populated town. Laki is, instead, a series of low cones in barren ground tucked out of sight from the sparse farm settlements of Iceland's southern coast. It is off the beaten track even for tourists making the rounds of the country's popular ring road.

Laki also falls short by other criteria. On the volcanic explosivity index – the standard 8-point scale scientists use to measure an eruption's power – Laki rates a modest 4, one notch below the 1980 eruption of Mount St. Helens in Washington. Laki did not erupt with the apocalyptic bombast of Indonesia's Krakatau, whose explosion in 1883 sent shock waves racing around the globe seven times, or Mount Pelée on the Caribbean island of Martinique, which in 1902 obliterated an entire town of 30,000 people within minutes. Laki did generate floods and fast-moving lava flows, but these generally followed the course of rivers and gorges and did not invade villages or settlements.

But Laki should not be underestimated because it didn't annihilate thousands of people outright. What made Laki particularly deadly – an insidious killer – was how long its eruption went on and what poisons it spat into the atmosphere. Over the course of eight months Laki produced one of the largest lava flows in historic times, enough to bury Manhattan 250 metres deep. Over the first twelve days it disgorged the equivalent of two Olympic swimming pools full of lava every second. Along with the lava came the gas: Laki belched out an estimated 122 million tonnes of sulphur dioxide, 15 million tonnes of fluorine, and 7 million tonnes of chlorine. It was one of the biggest atmospheric pollution events in the past 250 years.

Laki heaved and spewed particles that spread eastward across the North Atlantic and descended on Europe. All summer, people all over the continent choked on this caustic smog. As we will see in chapter 5, people could neither avoid nor escape

The Laki crater row stretches more than 27 kilometres and includes at least 130 separate craters, all created during the 1783-84 eruption.

the malignant haze, which manifested itself as a bluish or reddish 'dry fog' that smelled strongly of sulphur. It dimmed the sun and instilled panic across the continent. Londoners were used to pollution, but this volcanic fog was interminable and highly toxic. Those prone to respiratory or heart ailments were particularly vulnerable. Tens of thousands of people in England and France died of causes that were inexplicable to doctors and scientists at the time.

The vast cloud of Laki's sulphur emissions proceeded to encircle the northern part of the globe, reflecting sunlight back into space and cooling the planet by a degree Celsius for several years. The cooling trend wreaked havoc with the weather. In Europe and North America the winter of 1783–84 was one of the worst in 250 years. Crops perished, setting up long-lasting hunger and social unrest that some scholars have linked to the French revolution of 1789. Further afield, harvests failed in Egypt, India and Japan, leading to famines that killed millions.

Back in Iceland, perhaps the most lethal aspect of the eruption was how it poisoned the countryside. Laki was unusual in that it had tapped an underground magma chamber that was rich in fluorine. In low doses in humans, this element can have beneficial effects, such as the hardening of tooth enamel and the stabilizing of bone mineral. High doses, however, result in fluoride poisoning. For more than 7,000 square kilometres around Laki, the ground was heavily salted with fluorine. Animals grazing in polluted pastures ingested ash that corroded their intestines, leading to fatal haemorrhages. Others developed severe bone and teeth deformations. Death, when it finally came, would have been merciful.

With no sheep for meat or cattle for milk, and no fish to gather from poisoned rivers, the livelihoods of whole communities collapsed. Famine set in, slowly but steadily. Farmers who had worked the same land for generations abandoned their homes in a futile effort to get away from the killer volcano. Beggars and vagrants moved in. At least one man kicked out his wife for feeding too many of the hungry people who arrived at their farm's door; generations later, her descendant Haraldur Sigurdsson would become one of Iceland's most famous volcanologists. Rumours flew that officials in Copenhagen, from where the Danish king ruled Iceland, were planning to evacuate large swaths of the stricken country.

No wonder, then, that Laki is burned into Iceland's national psyche and taught in every history class. The eruption and its aftermath have become a benchmark for measuring all other painful episodes in that country. Icelanders have a word for it – *Móduhardindin* – meaning 'the hardship of the fog'. Laki was the most severe disaster to happen in Iceland for centuries, not just for the number of lives lost but also for its long-term effect on the development of the entire country. It helped keep Iceland as the remote place it had been for so long.

Although the Laki eruption would test Jón's faith and mettle almost beyond bearing, probably no one in Iceland was better prepared than he to cope with or understand the imminent catastrophe. By all accounts, he was a man of extraordinary aptitude and insight. He kept a near-daily chronicle of the months that followed 8 June 1783. This diary, known as the *Eldrit* or 'Fire Treatise', would not only document the terrible consequences that Laki would have for Icelanders, but also inspire future research into this exceptional eruption.

Jón was born in 1728, the eldest child of a farming family near the village of Hólar in northern Iceland. It's hard to overemphasize how simple rural life was at the time. Daily life focused on the health of a family's sheep, as one harsh winter or a livestock epidemic could reduce the strongest farmer to ruin. Social life revolved around the village; the king in far-off Denmark, which had ruled Iceland since 1660, was only a vague concept. Local officials appointed by Copenhagen oversaw such matters as markets and law courts, but otherwise the farmers handled most affairs among themselves, referring to the local preacher for guidance on occasion.

Like most such farmers, Jón's parents were pious and humble. His father taught him to bare his head and recite the Lord's Prayer and other words of blessing before tending the sheep, so that 'nothing evil would befall him'. During the day, he would regularly recite or hum hymns to himself. He even used one of his father's favourite hymns, 'One Lord I Prize Above All', to gauge distances. The hymn had ten verses, and it took him forty repetitions to get from home to the sheep pen.

By the age of nine, Jón had developed a keen interest and skill in working pasturage, a vocation he might have pursued had his father not taken ill and died the following year. His childhood came to a halt, and the family descended into near-poverty. Jón's mother, a widow with five mouths to feed, hired an overseer to supervise the labourers and manage the property. Somehow, the family managed to scrape by.

Young Jón hoped to rise out of impoverishment by attending the diocesan school in Hólar, which was intended mainly for the training of future priests. But his family could not afford it, and he spoke with a stammer that did not impress the teachers there. Eventually, with his mother's encouragement and support from others, Jón obtained a scholarship and proved to be an excellent student. He graduated in 1750 with a solid foundation in Latin and classical literature.

By this time, aged twenty-two, Jón had also developed certain characteristics that today we might regard as eccentric. As we have seen, along with his profound faith and an interest in nature he believed in portents and dreams as revelations of future events and God's will. He considered himself sensitive to ghosts, and in his autobiography he recounts several stories of encounters with evil spirits, including one in which a poltergeist lifted him bodily in the air and threw him across the room. Jón was not alone in his supernatural worldview, however. Many Icelanders believed (and still do, to some extent) in the existence of spirits, monsters and 'hidden people' or *huldufólk* (elves and trolls). Mystical drawings and incantations, it was thought, could harness supernatural powers for good or ill, and some individuals were held to be clairvoyant. Such convictions, along with strong religious beliefs, lent order to a world that was rife with inexplicable forces.

After Jón graduated, many of his friends and relatives assumed that he would sail to Denmark to continue studies at the University of Copenhagen, as most young promising Icelanders did. But providence had other plans.

At a farm near Hólar lived a wealthy farmer named Jón Vigfússon and his wife, Thórunn. Vigfússon, a former soldier, often went on drinking binges for days at a time, during which he might wave his sabre at anyone who crossed him. More often than not he turned his aggression on his wife, beating and choking her. If any man tried to stop him, Vigfússon would accuse him of sleeping with Thórunn.

Concerned, the bishop of Hólar offered to appoint Jón as deacon of the church that stood on Vigfússon's property, partly in the hope that his calming presence might quell the man's violence. Jón accepted the post, leaving dreams of Copenhagen behind.

For a while, Vigfússon's drinking and brutality continued unabated – the situation was so bad that on one particularly grim day, Jón and some members of the household agreed that 'it would be better if the master were dead'. Astonishingly, not long after, the master was found dead in his bed, apparently from liver or heart failure.

Vigfússon's decease was Jón's relief, because he was secretly in love with the abused wife. Thórunn was not a natural beauty. Life had been hard for her: in twelve years of arduous marriage she had borne five children, two of whom had died in infancy. As Jón wrote later, she was 'hollow-backed' with a 'protruding stomach' and 'a recessed hairline'. Smallpox had disfigured her face. 'She had a very striking and attractive appearance when seen from behind,' Jón wrote, seemingly without irony, 'and gave an impression of utter goodness – a family trait – when one saw her from the front.'

Once Vigfússon was gone, Thórunn and Jón moved in together. Nearly a year to the day after his death, they were married. A very inconvenient few months later, however, their first daughter was born. The church took note, and Jón was removed from his post. Thórunn, who was fairly well off, owned property in southern Iceland, and in 1755 the couple decided to make a fresh start and move their family there. In September of that year, just before Katla erupted, Jón set out across central Iceland with his brother and a hired man to prepare for the move. The volcano turned out to be the least of their concerns. During the months-long journey, they nearly died in blizzards and bitterly cold temperatures. Their perseverance finally paid off, and by the beginning of winter they had made it to the south, settling on a farm by the sea, near present-day Vík. The farmer said they could live in a storehouse on the property, which was little more than a cave hewn into rock. Nevertheless, the brothers lived there happily during the winter, and the family joined them in the following spring.

Over the next several years, Jón became a prosperous farmer and fisherman. This won him much admiration among his new neighbours, who often sought his counsel. He also caught the eye of the local priest, who, in failing health and unable to serve the entire district, invited Jón to become his assistant. Jón resisted at first because he and Thórunn were comfortable with their newly restored lives at the farm. More importantly, he was reluctant to serve under an 'inflammable priest' whom he thought had lax morals and was a heavy drinker. The priest, however, slyly solicited the help of Jón's friends to convince him to take the position. Not wishing to offend them, he relented. In the final agreement, it was decided that the priest and Jón would oversee separate parishes in the district. Jón was apparently popular as an assistant priest for, a year later, most of the farmers in the parish sent him letters asking him to be their permanent clergyman. Wishing to separate himself entirely from the inflammable priest, Jón sent these letters to the bishop, who agreed to the arrangement. Accordingly, in 1760, Jón was ordained to the priesthood and moved to Fell farm, the local rectory.

Priests at the time were expected to be farmers, and Jón's boyhood training served him in good stead in that respect. He also developed another skill that garnered even more admiration from his parishioners: medicine. Being the autodidact that he was, Jón had studied the medical arts on his own and began practicing at Fell. He also trained for a while with a physician, natural scientist and former schoolmate named Bjarni Pálsson, who went on to become Iceland's first surgeon general. Jón treated many people for free and was seldom without a resident patient in his home, sometimes accommodating two or three at a time. He later estimated that he successfully treated up to 2,000 people during his seventeen years of active practice.

Chalices, church bells and other sacred ornaments constituted the few fine possessions of Jón's spiritual flock.

In 1778, the position of priest for the Sída district, to the east, fell vacant. Jón applied for and won the appointment. Now fifty years old, he had enjoyed being a priest in Fell, but he was ready to move on. The most painful experience was a lawsuit brought against him and Thórunn by Jón Scheving, the son of Thórunn and her detested late husband. Scheving had squandered his inheritance and was living a deadbeat lifestyle in Copenhagen. For some time, he had apparently harboured an intense hatred for his stepfather and mother, and he devised a plan to ruin them and appropriate their money. He bribed a hired hand to claim that his mother and Jón Steingrímsson had conspired to have his father murdered so that they could live together unimpeded. News of this scandalous accusation spread rapidly through Copenhagen and across to Iceland. Jón's utterance that 'it would be better if the master were dead' came back to haunt him, because many people believed the stepson's story. The lawsuit unravelled, however, when the hired hand retracted his testimony under investigative pressure and admitted that Scheving had paid him to make the charge. Upon learning that his scheme had been thwarted, Scheving signed up with a regiment of soldiers and transferred out of Copenhagen. Jón never heard from him again.

Although Jón and Thórunn were exonerated, the experience continued to weigh heavily upon them, and when the position became available in the Sída district they jumped at the chance of moving. The job was in the town of Kirkjubæjarklaustur, commonly known as Klaustur, a tranquil place with a rich religious history. Well before the Víkings arrived to settle Iceland permanently, Irish monks occasionally visited the broad, richly forested slopes beneath the steep cliffs of volcanic rock. By the twelfth century nuns had established a famous convent here (the 'kirk' in the town's name), and many of the natural features in and around Klaustur are still named for the sisters. It was a place Jón could feel comfortable in and thrive, and he would serve it for the rest of his life.

Klaustur, it turned out, would also be an ideal location for witnessing the end of the world.

In the days following the first appearance of the black cloud, Jón recorded new and more inauspicious manifestations coming from the direction of Mount Laki. On 9 June, the day after Whitsun, the weather started out clear and sunny, but ominously the dark cloud returned and was rising ever higher in the north. Earthquakes were drumming faster and faster, accompanied by thuds and loud cracking sounds. The Skaftá River, which normally flowed at a volume so great that horses at the ferry crossing had to swim some 120 metres through the current to ford it, suddenly began to drop. An acid rain fell the next day, eating through pigweed leaves and scorching the hides of newly shorn sheep. By the afternoon, the Skaftá had dried up completely. Snow fell from the black cloud on 11 June, creating a hard, shellac-like surface over the grass. The sun, when it could be seen, was red as an ember, and the moon blood-coloured.

Then, on 12 June, lava gushed forth 'with frightening speed' from the Skaftá canyon southwest of Klaustur. As it encountered wetlands and other streams feeding into the river, the combination of water and molten rock created concussive explosions. At first, the lava followed along the course of the riverbed, but soon it breached the banks and began spreading over old lava flows, meadows and farmlands. This was to be the first of five such surges of lava from the gorge.

Cinders fell from the sky on the 14th. They were, Jón wrote, 'blue-black and shiny, as long and thick around as a seal's hair'. This was the first description of what volcanologists today refer to as Pele's hair – thin strands of volcanic glass formed by molten particles ejected in a lava fountain and stretched into fibres as they are carried through the air. Just half a millimetre across, they may be as long as two metres. The unusual fallout covered the ground across the region, and the winds worked some of the hairs into long hollow coils.

Friction among ash particles generates ominous-looking lightning in volcanic plumes, as in the 2010 eruption of Eyjafjallajökull.

The following day, a party of farmers decided to climb Mount Kaldbakur, eight kilometres northeast of Klaustur, to see if they could get a good view of the eruption site. They reported seeing lava coursing through the river gorge, and, off in the distance, twenty fountains of fire exploding high into the sky. The news terrified everyone, including Jón, for it seemed certain now that the lava would breach the mountains and ravage the settlements.

Throughout the waning days of June, the forefront of the lava flow turned southeast, engulfing meadows and woodlands and laying waste to farms and churches. Birds fell dead from the sky and fish floated lifeless to the surface of streams and ponds. Earthquakes convulsed the ground and acrid odours filled the air, along with smoke and ash so thick that no one dared inhale deeply. The water tasted of sulphur and freshwater pools were fouled with ash. Thunderous eruptions could be heard coming from the glaciers up in the mountains. At night great showers of sparks shot into the sky. Lightning produced in the ash clouds was violent and at times continuous, so that 'scarcely a moment passed between bolts for days on end.' By early July, new lava was seen flowing under older lava, creating convulsions of subterranean fire and steam that caused the earth to heave upward and crack open.

Some people desperately tried to relocate their livestock, but their efforts usually came to naught. The proprietor of one of the region's most prosperous farms collected 'a great number' of his sheep and placed them on a small island in a river, intending to herd them to safety as soon as he had an opportunity. Before he had returned to his house, however, the lava came rushing on more quickly than he expected. It rapidly engulfed the island and consumed the sheep.

Many owners abandoned their homes and land, vowing never to return. Others made preparations to leave and then decided to wait and see if the lava flow would reach the sea, in which case they and their families would flee eastward. 'All the schemes, projects, and remedies that people undertook,' Jón wrote, 'led to confusion, frustration, exhaustion and expense, and in most cases were totally unavailing.'

Between July 13th and 19th, the lava edged further down the Skaftá riverbed toward the east, making its way – seemingly inexorably – toward Klaustur and Jón's church. In some places the lava piled up so high that it blocked the noonday sun. In other places floodwaters inundated farmlands, reducing them to muddy sloughs. For more than a week, suffocating clouds of smoke blanketed the area, forcing people into their homes. Because of the encroaching lava, the estate overseer at Klaustur decided to remove as many valuables and ornaments from the church and cloister as possible. For Jón, relocating church accessories such as the altar, chalice, paten and other sacred vessels must have been a sorrowful turning point.

By 17 July, residents fleeing the lava's advance west of Klaustur were streaming into the area, herding their cows before them. (Most of the terrorized sheep had fled in all directions.) Two nights later, the tumult subsided somewhat, though the relative calm was occasionally broken by thunder and distant cracking sounds. The approaching lava now lay less than three kilometres from the church.

In his home nearby, Jón lay in sleepless dread, praying and fretting over a terrible, dawning truth: that tomorrow would most likely be the last day he would ever hold service in his beloved chapel. Its destruction seemed certain.

CHAPTER TWO

Land of Ice and Fire

The volcanoes of Iceland

ICELAND IS AN ISLAND OF DESTRUCTIVE FIRE. Jules Verne sensed this when, never having visited the place, he conceived his novel *A Journey to the Centre of the Earth*. In many ways Verne got the basics of Victorian-era geology wrong: his protagonists encounter the most unlikely wonders, including a cave filled with giant mushrooms and a battle between prehistoric sea beasts. But in at least one aspect Verne was both accurate and far ahead of his time. To reach the centre of the Earth, he sent his characters down the throat of Snefels (Snæfellsjökull) volcano, 120 kilometres northwest of Reykjavík. This mountain, Verne correctly suspected, was a direct conduit to the planet's internal fire.

Almost since the day Iceland was settled in the ninth century, explorers, naturalists and writers have viewed the island as a geological wonderland. Here were full-throated volcanoes spewing fire and ash, with bubbling hot springs and geysers spouting wildly into the air. Here, too, were great ice-mountains, or glaciers, which ground their way down from lava-capped peaks. Iceland combined the dramatic flames of Italy's volcanoes, such as Sicily's mighty Mount Etna, with the snowy grandeur of the Swiss Alps. It's little wonder, then, that the island quickly garnered a reputation as a marvellous land of fire and ice. Iceland is like nowhere else on the planet.

To understand why this should be, we have to take our own virtual journey to the centre of the Earth. By listening to how seismic waves bounce through the planet's interior, geophysicists have worked out that the Earth is made of three main layers, each with its own characteristics. At the surface is a thin outer crust, like the skin of an apple; everything people see and do in their everyday lives happens on this crust. Beneath that, stretching from a depth of around 40 kilometres to 2,900 kilometres, lies the mantle, which makes up most of the planet, like the flesh of the apple. Finally, beneath the mantle beats the planetary heart: a spinning core made partly of liquid iron, whose sloshing generates the Earth's magnetic field.

Volcanoes, in Iceland and elsewhere, are born hundreds of kilometres deep in the mantle. Most of the time, the crushing pressure at these great depths ensures that rock remains solid. The question is: what causes some of that rock to melt, rise toward the surface, and occasionally erupt as a volcano? People have been puzzling over this since at least the fifth century BCE, when the Greek philosopher Anaxagoras suggested that winds were forced among cracks deep in the Earth, heating up because of friction and melting the rocks around them. Five centuries later, the Roman philosopher Seneca proposed that burning fossil fuels, such as coal, fed the fire of volcanoes. Today, after 2,000 years of study and experiments, scientists appear to have a more precise answer: heat, chemistry and pressure all play crucial roles in causing mantle rocks to melt, rise and erupt.

Heat has been lurking in the planet's interior ever since it formed 4.5 billion years ago, when Earth coalesced from a disc of gas and dust swirling around the newborn sun. Much of the planet's internal heat comes from radioactive elements left over from its fiery birth. Such elements include thorium, uranium and potassium, which can take billions of years to decay into less radioactive substances. The nuclear breakdown releases heat that percolates up toward the surface.

Chemistry comes into play in determining which particular rocks are destined to melt. Most rocks are made of minerals in which atoms of elements such as silicon and oxygen are arranged in regularly repeating, or crystalline, patterns. Other elements, such as iron or magnesium, tuck themselves into this arrangement. Depending on what atoms are in the structure and how tightly they bond to one another, the atoms can be snuggled close together or strung out more loosely, like pearls on a string. When temperatures rise, the first chemical bonds to break are those between atoms that are less tightly linked; thus different minerals have different melting points, depending on how their chemistry is arranged. A rock made up of different minerals will melt its mineral types one by one rather than all at once. Over time, the chemistry of the rock changes as part of it melts out while the rest remains solid.

Finally, pressure is the key that links heat and chemistry together to produce molten rock. If you take a chunk of surface rock and raise its temperature to the melting point, it melts, naturally. But put that same chunk hundreds of kilometres deep in the mantle, and the high pressures will keep it solid. That's why most of the Earth's mantle is solid. It's hot, to be sure, but the pressure is high enough in most places to keep the rocks from melting completely.

As a liquid, magma is less dense than the surrounding rock, and so it starts to rise. Magma will take any path it can: it might worm its way through pre-existing fractures in the rock, or shoulder its own path upward. Often flows will pool together in a shallow reservoir in the crust. Such holding chambers lie beneath most of the world's volcanoes, and are the reason why a mountain considered dormant can come to life again at any time.

Once magma reaches the surface, the eruption can take any number of forms. It might shred the magma explosively into a plume of tiny ash fragments jetted kilometres into the air – as Iceland's Eyjafjallajökull did in 2010, shutting down airspace across Europe. Or the volcano might ooze lava, so that glowing currents of rock course down the sides of the mountain – like Hawaii's Kilauea, which frequently and picturesquely sends rivers of lava into the sea. Alternatively the eruption might be some combination of the two, belching out both ash clouds and lava flows at different times. But no matter what the volcano looks like at the surface, its magma always comes from melting deep within the Earth.

Which is why volcanoes don't just appear willy-nilly across the planet, but instead crop up where geological forces trigger and concentrate melting – most famously along the Pacific Ocean's Ring of Fire, linking Indonesia to Japan to Alaska to the Andes. And for that we can thank the phenomenon of plate tectonics.

The science of plate tectonics has a deep history, but it crystallized in the work of a German explorer and meteorologist named Alfred Wegener. In 1915, while recovering from war injuries, Wegener wrote a book laying out his ideas. Like many before him, including almost every child playing with a globe, he noted that the east coast of South America could nestle neatly against the west coast of Africa like jigsaw puzzle pieces fitting together. He traced the span of the Appalachian Mountains along a map and thought they linked up with other ranges across Ireland, Scotland and Scandinavia. He also noted that fossils of many species in Europe and North America looked fairly alike until around 180 million years ago, when they diverged dramatically. To Wegener, this suggested that something must have pushed the landmasses around over time.

But how could the Earth's seemingly solid crust move like that? Wegener thought the answer lay in the difference between the land and the oceans. He noted that most of the continents lay between sea level and an altitude of 1,000 metres, whereas

Alfred Wegener (left) and his guide Rasmus Villumsen on their ill-fated final expedition to Greenland's ice sheet. Both men died on the 1930 trip.

the oceans were mostly between 4,000 and 5,000 metres deep. For this arrangement to work, Wegener argued, the continents must be made of lighter material that floats on the denser ocean crust below. That would keep mountains at relatively modest heights compared to the oceans' abyssal depths. More importantly, it suggested that continents and oceans were separate things, and that each behaved in its own way. Continents could drift around on this ocean base, like icebergs sailing through the sea.

Wegener didn't have much luck convincing other scientists of this 'continental drift' theory, mainly because he couldn't explain what might start continents moving in the first place. And when he died on an expedition to Greenland's ice sheet in 1930, still seeking to prove his idea, much of the impetus to understand the concept died with him.

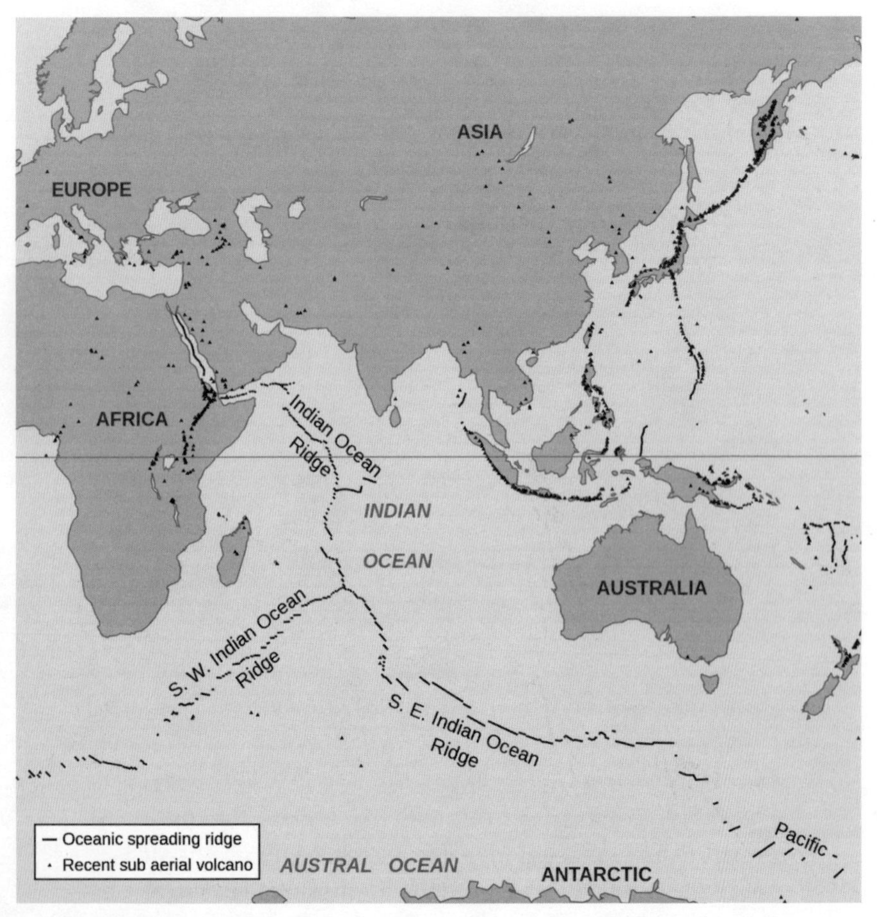

Mid-ocean ridges run down the centres of oceans like seams on a baseball.

Until 1962, that is. That year, geologist Harry Hess of Princeton revived Wegener's ideas. As a young researcher Hess had sailed on oceanographic expeditions to map gravity anomalies in several ocean trenches; from these he began formulating ideas about the rates at which oceanic plates might be moving. In 1962 he published what he called an 'essay in geopoetry', written two years earlier. It proposed that heat churning through the Earth's mantle might be responsible for the movement of the plates. And the essay reopened the question of Wegener's continental drift.

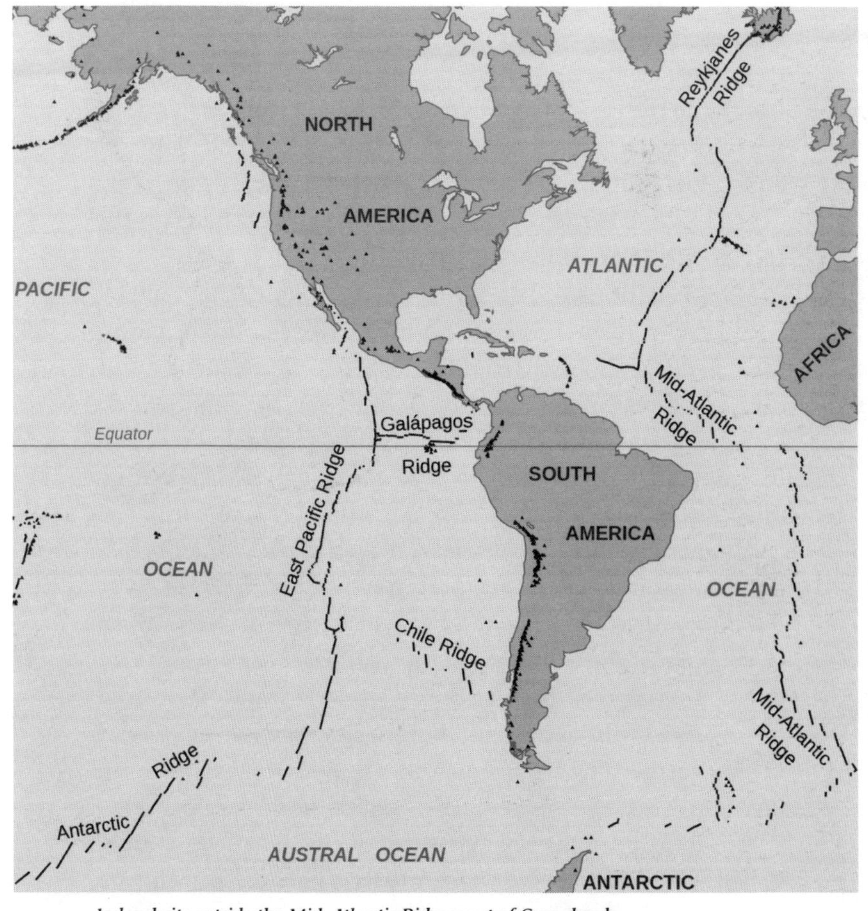

Iceland sits astride the Mid-Atlantic Ridge, east of Greenland.

Hess proposed that it all starts in the middle of the oceans, where lava spills out along great gashes in the sea floor. The lava cools as it hits the frigid seawater, and piles up as underwater mountain ranges. Picture a globe with the water drained away, and these 'mid-ocean ridges' snake down the centres of the oceans like seams on a baseball. If laid end to end, they would stretch for 50,000 kilometres or more. They are the biggest single volcanic feature on the planet.

Hess argued that fresh ocean crust is made at mid-ocean ridges every day. Heat churning within the mantle, rising up like the blobs within a lava lamp, carries the newborn seafloor away from either side of the fiery seam. More magma then wells up from the deep, bubbling up and hardening into new rock to fill the gap. The new crust rides away from each side of the mid-ocean ridge, as on a conveyor belt. The belt's journey ends, Hess proposed, when the rocks reach the far side of the ocean basin, where they jam up against and dive beneath the higher-floating continental crust.

His theory described an ocean bottom in which the youngest ocean crust is closest to the mid-ocean ridges, and the oldest crust is that smashing into the continents. Conveniently, this idea could readily be tested by oceanographers, and several 1960s expeditions did just that, towing magnetic detection instruments behind ships to read the magnetic orientation of rocks on the seafloor.

Magnetic studies can reveal the ages of rocks thanks to a peculiar habit of Earth's magnetic field: it flips direction every few hundred thousand years, or at even longer intervals. When this happens, the north magnetic pole essentially turns into the south magnetic pole, and vice versa. If you were around during one of these flips and holding a compass, you would see its needle turn from north to south. It might, however, take you 1,000 or 10,000 years to watch the entire flip.

The seafloor rocks hold a record of this magnetic switch. When lava cools, its iron-rich crystals become frozen in a particular orientation, pointing toward magnetic north like a forest of tiny compasses. The last magnetic field reversal took place 780,000 years ago, so any rocks that cooled in the last 780,000 years would have compasses pointing toward what we know as north today. The period before that, when rocks would cool with their compasses pointing south, lasted between about 950,000 and 780,000 years ago.

As seafloor rocks form and move away from the mid-ocean ridge, then, they should preserve a record of this alternating magnetic polarity through time. And that's exactly what the oceanographers found when they drilled and retrieved crust from the seafloor: stripe after stripe of normal magnetism, then reverse magnetism, then normal magnetism. It all stretched away from the seafloor spreading centre, symmetrically on each side of the mid-ocean ridge, like a huge tape-recording machine capturing more than 100 million years of seafloor history.

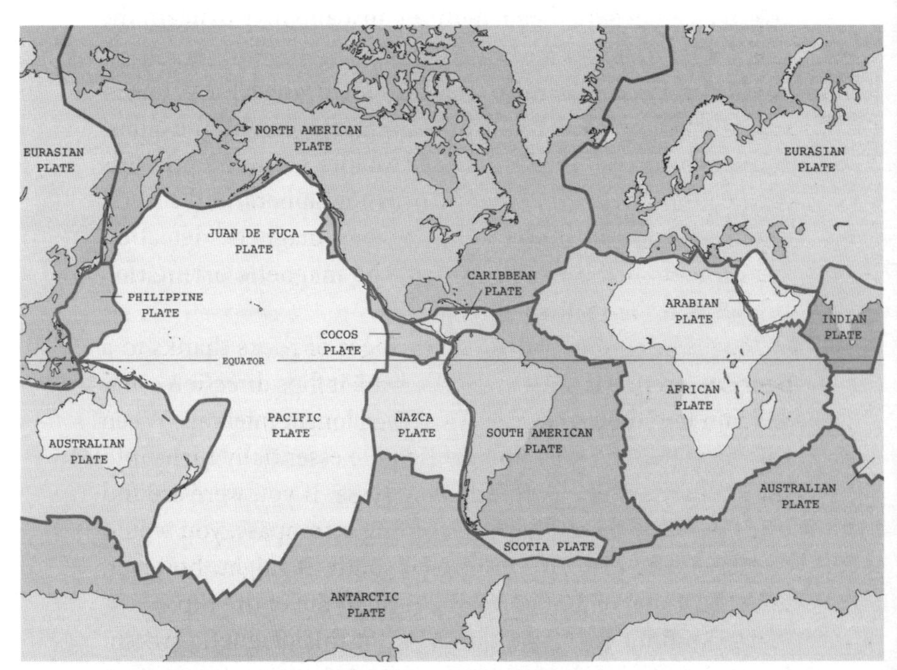

The Earth's crust is divided into about a dozen major tectonic plates, plus a clutch of smaller ones.

With this confirmation of seafloor spreading, geologists had the full picture of plate tectonics. Volcanic activity from below created the mid-ocean ridges. Fresh crust moved away on each side of the ridge, slowly and steadily, until it smashed into a continent. The oceanic crust, being denser than the continent, plunged beneath it, in a process that geologists call subduction. Finally, scientists had both the evidence that continents moved, and – the biggest piece that Wegener never got – the reason why they did.

Today, researchers know that Earth's crust is divided into about a dozen large tectonic plates and many smaller ones, which jostle against one another like schoolkids elbowing for a place at the lunch table. Over hundreds of millions of years the physical pressure of plates slamming into one another, fuelled by heat from below, have rearranged the oceans and continents. The supercontinent Pangaea ruled the planet around 250 million years ago. It later fragmented into two smaller landmasses and then into the seven continents we know today. Scientists have even calculated how plate tectonics might shift continents in the future: in the next 200 million years, because the Pacific seafloor is being subducted, North America and Asia will merge to form a supercontinent, which has been dubbed Amasia.

Plate tectonics dictate most of the planet's geological activity. Where plates meet, huge geologic disasters can happen. Zones of plate subduction, such as where Pacific oceanic crust dives beneath South America or Japan, are often the sites of Earth's greatest earthquakes. The part of the plate diving down usually sticks, building up stress and then releasing it in periodic spasms – like the magnitude 9.0 earthquake (and resulting tsunami) that devastated Japan in March 2011.

At the same time, the subducting ocean plate also leads to volcanoes. The plunging plate carries with it sediments from the ocean floor that contain trapped water, which causes the rock to melt at a lower temperature than it otherwise would, so that the ocean slab and the sediments on top begin to melt

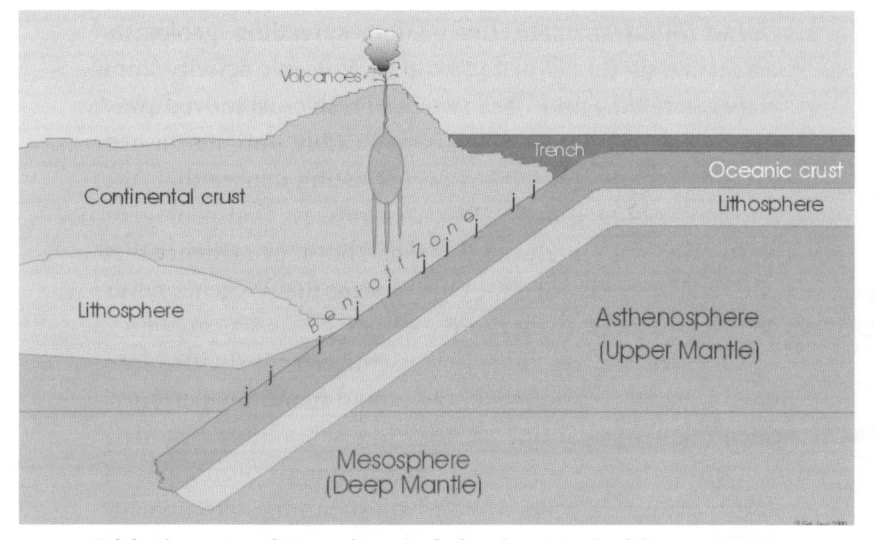

Subducting ocean plates: a downward-plunging tectonic plate can generate volcanoes in the overlying crust.

fairly soon after diving beneath the continent. This melting feeds new magma chambers which, in turn, feed volcanoes above them. This is how the volcanoes of the Pacific Northwest, like Mount St. Helens and the rest of the Cascades, are made: part of the old Pacific ocean plate is plunging beneath North America, and molten material is rising to fuel the Cascades. On a grander scale this is also why the entire Ring of Fire exists, encircling the Pacific with a chain of active volcanoes. Scientists nicknamed this chain long before they understood the plate tectonics that explained its existence.

Most of the world's volcanoes, then, can trace their origin either to new ocean crust being created at mid-ocean ridges, or to subduction zones. But some iconic volcanoes sit far from these violent collisions of tectonic plates. Consider the Hawaiian islands, for instance. They sit in the middle of the Pacific, but far from any mid-ocean ridge, and experience essentially no earthquakes other than the small ones caused by the Hawaiian volcanoes themselves. Part of the explanation for Hawaii's existence lies with another geological wonder:

plumes of hot rock that rise all the way from the mantle, like a welder's torch switched permanently to 'blast'. In a mantle plume, a jet of volcanic heat fuels a constant, massive outpouring of lava on the seafloor. Scientists think that a few dozen mantle plumes dot the planet, of which the Hawaiian islands are the most famous example.

The Hawaiian islands get progressively older the further northwest you go. Kilauea, the currently erupting volcano, is on the active Big Island in the southeast. The scenery here is an eerie moonscape of black-as-night lava flows. Go northwest, through Oahu and Maui, and the travellers' paradise gets progressively lusher. By the time you get to Kauai in the far northwest the volcanoes haven't been active for six million years. That's why Kauai's landscape is so verdant and full of coffee and fruit plantations: it has had plenty of time for its lava to erode into a soil that can nurture plant life.

Hawaii can be explained by a combination of plate tectonics and mantle plumes. Over time, as the Pacific crustal plate drifted over the stationary mantle plume beneath, the blowtorch heat created one island after another, embossed like Braille dots on a moving sheet of paper. Follow the chain even further to the north and west, and you'll encounter Midway Island at 28 million years old, and then a sharp kink, with volcanoes around 43 million years old, where the underwater Emperor seamounts take over, at ages regularly stair-stepping up to 65 million years. This bend marks a time when the Pacific plate must have taken a sharp turn and moved in a different direction over the mantle plume. It is geological history written as islands.

In Iceland, though, mantle plumes and mid-ocean ridges come together. Beneath the island lies a plume that has been active for the past 65 million years. It has poured out some 10 million cubic kilometres of magma – 50 times the volume of Iceland itself – across the North Atlantic. Traces of its activity range from Scotland to Greenland.

This prodigious plume helped the Atlantic to split where it did. As the Eurasian and North American tectonic plates pulled apart, about 25 million years ago, the mid-Atlantic ridge locked onto the mantle plume rising from below. Plume and ridge have been delivering a double dose of magma to Iceland ever since. Geologically speaking, the island is but a babe. If Earth's history were compressed into a year, with the planet being born in the first second of 1 January, then Iceland wouldn't have shown up until the afternoon of 30 December. It is still growing, as eruptions add more fresh land to the island than is lost to the sea. Today the centre of the mantle plume lies beneath the northwestern part of the Vatnajökull icecap, to the north and east of Laki. The plume continues to build the island and fuel its volcanoes.

It also dictates how Iceland pulls itself apart. As the mantle plume remains fixed in place, the mid-ocean spreading centre keeps moving to the northwest, by fractions of a centimetre per year. The plume works to adjust the crustal stresses by creating new rifts. A map of Iceland reveals these geologic forces battling one another: the mid-Atlantic rift runs up from the southwest of the island, bends toward the east along its centre, and finally turns northward again and exits the island. Each of these rift segments has its own collection of volcanoes and earthquakes.

In Iceland, volcanism is exposed in all its raw glory. You can observe the island's geological turmoil at Thingvellir National Park, about an hour's drive east of Reykjavík. Just past the parking area for tour buses on the popular 'Golden Circle' day trip, the mid-Atlantic rift appears as giant ridges of fresh lava. You can walk along the rift here; a paved pathway takes you along the base of a huge lava cliff, which was raised as the land split apart. To your east, the Eurasian tectonic plate is moving away from you; to your west, the North American plate is doing the same thing. From time to time, the tourist pathway gets destroyed by the ongoing rifting, which happens at about the rate that fingernails grow.

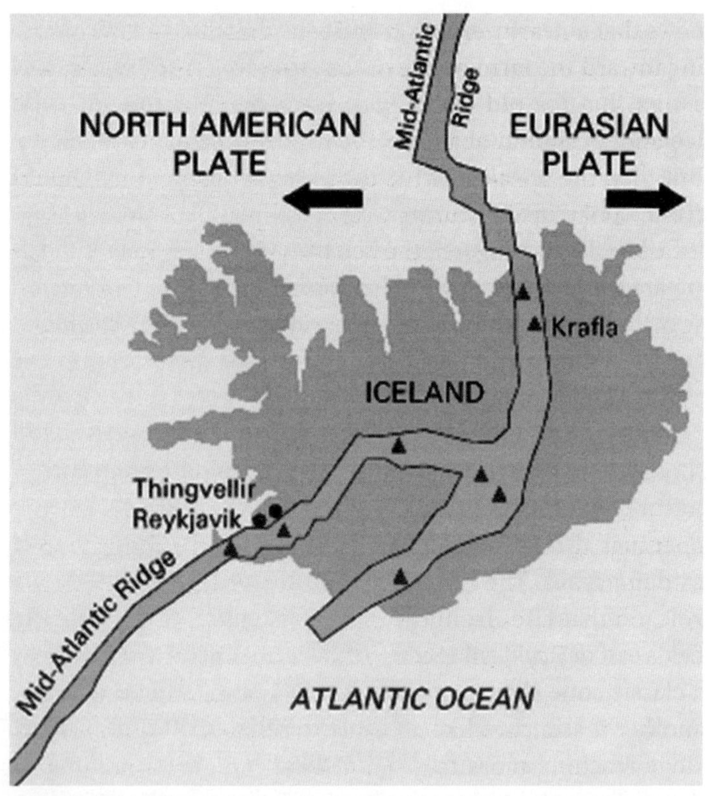

The Mid-Atlantic Ridge splits Iceland in half, with North America on the west and Eurasia on the east.

This dramatic setting made Thingvellir an appropriate site to hold the world's first democratic parliament, the Althing, a millennium ago. Chiefs from various parts of Iceland regularly travelled to Thingvellir to discuss topics of common interest and come to general decisions. At any given time the designated speaker stood at the base of the cliff, a spot that amplified his voice and spread it over the gathered crowds. In the year 1000, this natural megaphone came in handy when the Althing began debating whether Iceland should convert to Christianity.

What happened next is a national legend. As the chieftains argued among themselves, a messenger appeared with the news that a nearby eruption had sent a fresh lava flow coursing toward the farm of one of the attendees. This, surely, was a sign that the old Norse gods were angry at the idea that Icelanders should abandon them in favour of Christianity. But then the Speaker of the gathering, a man named Snorri Godi, spoke up. Gesturing across the plains of the Althing, he asked: 'What angered the gods when the lava burnt which we are standing on now?' After a moment for the message to sink in – that eruptions were a natural part of life in Iceland – everyone concurred. The Althing voted to adopt Christianity as the island's religion.

With Christian theology came such concepts as Hell. For those looking to instil a fear of an afterlife of subterranean torture, Iceland's fiery mountains provided a powerful aid: the openings through which lava poured became the gateways to damnation. The main gateway was Hekla, the immense volcano that rises from the lava plains near the great geyser fields east of Reykjavík. Seen from one direction Hekla assumes a classic cone shape, an icon of volcanic majesty; seen from another it stretches like an overturned boat, a snow-capped ridge reaching more than 1,490 metres high. Its name may derive from the Norse word for 'cloak', after the mists that often enshroud the mighty mountain.

Travellers from abroad soon began journeying to Iceland to see this otherworldly portal, through which arose the sounds of souls being tortured. As early as the twelfth century, visitors from continental Europe marvelled at its fury; one French cleric reported that Sicily's famed Mount Etna was 'like a small furnace compared to this enormous inferno.' Lest the main message be lost, the monk added:

> *Who now is there so refractory and unbelieving that he will not credit the existence of an eternal fire where souls suffer, when with his own eyes he sees the fire of which I have spoken? ... But whosoever will not believe in or hear spoken of the punishments of eternal fire that are prepared for the*

Devil and his friends, he will later be hurled into that place which he cares not to avoid while he can.

In the sixteenth century, the German physician and traveller Caspar Peucer continued the theme: 'Out of the bottomless abyss of Heklafell, or rather out of Hell itself, rise melancholy cries and loud wailings, so that these lamentations may be heard for many miles around . . . Whenever great battles are fought or there is bloody carnage somewhere on the globe, then there may be heard in the mountain fearful howlings, weeping and gnashing of teeth.' In the seventeenth century, a travel book described how the devil would pull souls out of Hekla and throw them onto the sea ice, cooling them off temporarily in order to exacerbate their torture when tossed back into the inferno. And when English poet William Blake needed somewhere to banish the spectre of winter, he did so in Iceland: 'The mariner cries in vain . . . till heaven smiles, and the monster is driv'n yelling to his caves beneath mount Hecla.'

Most Icelanders, however, never went in much for the concept of Hekla as Hell's doorstep – they regarded the mountain's fire as just another natural phenomenon of their extraordinary island. In 1540, an Icelandic cleric visited the Danish king in Copenhagen and wondered why he was asking so many 'unneedful' questions about Hekla. Some Icelanders began a public-relations campaign to fight back against the notion of their island nation as an inferno for the damned. A bishop named Gudbrandur commissioned a scholar to write booklets to be published in Europe, correcting what he saw as misinterpretations of Iceland. It didn't work: authors kept on associating the country with the flames of perdition. But the scholar who wrote the booklets, Arngrímur Jónsson, was inspired to write an influential history of Iceland, the *Crymogaea*, which served to alert Europe to the fact that the country had its own cultural heritage and was not just some lava-scorched backwater.

Still, the combination of volcanic flames and Norse heritage was too powerful not to influence Icelandic mythology. The thirteenth-century *Poetic Edda*, one of the foundation texts of Scandinavian literature, includes a description of the sky darkening so that the sun and stars disappear, as smoke and flames rise toward heaven. Some scholars have argued that all Scandinavian myths of an apocalyptic conflagration, such as the great battle of the gods at Ragnarök, draw heavily on observations of eruptions in Iceland. Researchers have even linked the very concept of Ragnarök to specific eruptions of the Bardabunga volcano, beneath the Vatnajökull ice cap, or to Katla, below a smaller ice sheet to the west.

The undisputed queen of Icelandic volcanoes is still Hekla, which has erupted at least twenty times since the island was settled. As far as Iceland's new residents were concerned, the mountain first rumbled to life in 1104. This 'first fire in Hekla' blasted rock fragments across much of the northern part of the country, making it the second biggest eruption in Iceland's recorded history. But the volcano has done even worse things in the distant past, such as burying nearly the entire island in ash around 2,900 years ago. By digging trenches to study the build-up from past eruptions, geologists have been able to work out a detailed chronology for how often Hekla has blown, and when. (In Reykjavík shops you can buy a cube of notebook paper with layers in different colours, each representing a different Hekla ash layer from the time of settlement to the modern era.)

Time and again, Hekla showered nearby farms with ash, rock and other debris. Carbon dioxide from eruptions crept down the sides of the volcano and collected in nearby hollows, smothering livestock until farmers could dig ditches to drain away the deadly gas. Further from the mountain, fluorine-laden ash settled onto fields, where sheep grazed on the poisoned grass and died from the toxins they had ingested. Through

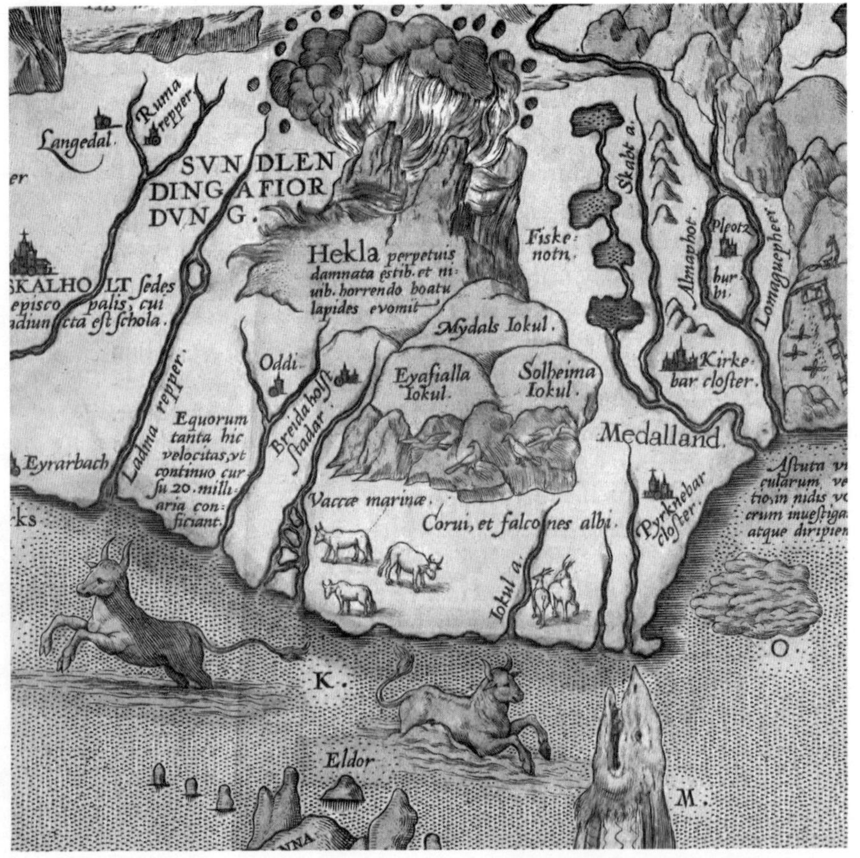

Abraham Ortelius's map of Iceland, seen here in a 1612 engraving, depicts the volcano Hekla in eruption. It was a popular cartographic reference.

it all, Icelanders continued to regard Hekla as a troublesome neighbour, whose occasional outbursts were to be understood and worked around rather than feared.

A chronicle of the mountain's eruption in 1300 C.E. describes people being unable to make their way across the countryside or put to sea in boats because of the 'sand' that blackened the skies and covered the ground all around. In 1845, Hekla sent a finger of lava down a gully toward its northwest flank, cutting off the water supply to a farm that had stood closest to the volcano for almost as long as there had been settlers in the island. Rather than panic, the farmers pragmatically tore down their barns and homes and rebuilt them on the other side of the lava flow.

Then, for the first time in the era of modern volcanology, Hekla rumbled to life in the early morning of 29 March 1947. Local farmers felt their bedrooms shaking and saw a brown cloud rising from the cloud-covered summit. Within hours, volcanologists packed a plane and left from Reykjavík to scout out the eruption. 'No words can describe the awful grandeur of the sight which there met our eyes,' wrote the university's Sigurdur Thorarinsson in his book *Hekla on Fire*:

> *From the northeastern part of the ridge there rushed forth dark, greyish-brown eruption columns, monstrously big around and so closely set that they formed a solid wall. In swirling puffs resembling gigantic heads of cauliflower these columns rose to a height of 10,000 metres. In the upper reaches their colour turned a lighter grey, and at the top they showed almost white against the blue sky. From time to time the fiery glow of the fissure showed through, and occasionally lightning flashed inside the cloud.*

The 1947 eruption spewed around a cubic kilometre of material across the landscape. Had Hekla erupted at this rate once a century since Iceland was born, it would have built the island six times over by itself.

Hekla's most recent eruption, in 2000, lasted only eleven days and did little damage. But Icelanders know their 'old lady' will undoubtedly go off again. In the meantime, they watch, wait, and wonder if it will be Hekla that erupts next – or one of the island's many other volcanoes. One strong contender is Katla, the country's second most active volcano, which lies about 50 kilometres to the southeast. If Hekla is the gateway to Hell, then Katla is the very devil.

Unlike Hekla, which is a barren mountain capped with only a little snow, Katla lies buried beneath hundreds of metres

Katla's 1918 eruption carried massive icebergs (note people for scale) along on its glacial flood, all the way to the south coast and the sea.

of solid ice. Whenever Katla erupts, it melts that ice from below. And that means lots and lots of water. Katla is one of Iceland's biggest sources of glacial outburst floods, in which water surges down the mountain and washes out any bridges, roads and farms that get in the way. Such sudden floods, called jökulhlaups ('glacier bursts'), are often the only indication that an ice-buried volcano has erupted. Jökulhlaups are the reason why many people call Katla the country's most dangerous volcano.

Folklore holds that Katla is named after a cook at a local monastery, who was lucky enough to own a pair of magic trousers in which a person could run forever. When a shepherd used these without permission one day to help bring in his sheep, Katla drowned him in a barrel full of whey. Eventually, realizing she would be caught for the murder, Katla put on the trousers, ran up the mountain, and threw herself into an icy crevasse. Soon afterward an eruption sent a glacial flood coursing downhill toward the monastery: Katla's revenge.

Katla's last eruption, in 1918, was the most powerful of any in twentieth-century Iceland. It sent jökulhlaups racing across the plains at speeds up to 20 kilometres an hour: witnesses reported that it looked as if the snow-covered hills themselves were moving, as house-sized chunks washed down from the top of the ice cap. When the waters subsided, icebergs as tall as eighty metres were beached on the coastal plain. The rushing waters left behind enough new land to temporarily claim the status of Iceland's southernmost point, until coastal erosion took much of it away.

Since then Katla has been quiet, other than possible small eruptions in 1955 and 1999. The gap – one of the longest in the mountain's recorded history – troubles many volcanologists. Especially because Katla has often erupted in tandem with another volcano 25 kilometres to the west, a once-obscure mountain that lurched to international fame in 2010.

The story of Eyjafjallajökull began uneventfully with what Icelanders call a 'tourist' eruption, one that's scenic but not very dangerous. In March 2010, lava began spewing from a fissure on the mountain's side, along a barren ridge that serves as a popular day hike. Photographers flocked to capture the fire fountains against the northern lights, and by volcanic standards it looked fairly unthreatening. It seemed as if the most problematic aspect of the eruption – for non-Icelanders, anyway – would be how to pronounce the name of the mountain. ('Eyjafjallajökull' means 'the glacier [jökull] atop the mountain [fjall] overlooking the islands [eyja],' and a YouTube search will turn up an Icelandic woman singing a ukulele ditty to teach people the word.)

But in early April, lava stopped coming out of the barren ridge. Something shifted in the plumbing system beneath the volcano and the eruption moved about a few kilometres to the west, directly beneath the glacier that caps Eyjafjallajökull. Suddenly, the magma got much stickier as it began to mix with other, older rocks in the magma chamber. It was also running into ice instead of air. The heat melted the ice from below, and then reacted explosively with the newly formed water. Now, instead of burbling out peacefully in scenic fire fountains, the magma shattered into ash fragments carried sky-high. Eyjafjallajökull was no longer a tourist eruption, but had awoken for real.

Winds happened to be drifting in just the right direction to ferry the ash toward the British Isles, and so the London centre in charge of volcanic-ash alerts told authorities to start shutting down airspace. That's because planes and volcanic ash do not play well together. When ash fragments get sucked into the extreme heat of jet engines, they can melt, forming a glassy coating that clogs up the turbines. In 1982, a British Airways flight in Indonesia passed through an ash cloud from Mount Galunggung and lost power in all four of its engines. So too did a KLM flight over Alaska in 1989. In both cases pilots eventually restarted the engines and landed safely, but not before dropping more than 3,000 metres. In the wake of these incidents, aviation authorities drew up new rules that forbid planes from flying in any airspace that could potentially have any amount of volcanic ash in it.

When Eyjafjallajökull erupted, officials had to abide by these rules. Every day, the London-based volcanic ash advisory centre issued maps of where the eruptive plume had drifted, and where it might go next. Planes could not enter any of that airspace. For nearly a week flights were grounded across Europe, stranding travellers and disrupting the flow of global commerce. The shutdown may have cost businesses as much as five billion euros, and carriers were told they must shell out for hotel expenses for their flightless passengers. Eventually, in reaction to the inevitable complaints, aviation authorities raised the acceptable limit of volcanic ash that planes can fly through.

The larger question is whether the annoying Eyjafjallajökull might set off the far more deadly Katla. Although seemingly

A May 2010 satellite image of Eyjafjallajökull shows its ash plume heading south and east toward Europe.

Heat from the buried Grímsvötn volcano causes lakes to pond atop the Vatnajökull ice cap (seen here in August 2011, a few months after an eruption).

separate on the surface, the two volcanoes may be linked by stress changes deep in the earth. Katla has regularly erupted on its own, but the last three times when Eyjafjallajökull erupted prior to 2010 – in 1821, 1612 and possibly the year 920 – Katla went off soon after. The connections between the two volcanoes aren't clear, and it's entirely possible that there is no real link between them, but volcanologists do think Katla could erupt at any time, and scientists and emergency planners are maintaining a very close eye on it. As if to keep everyone on their toes, in the summer of 2011 seismic activity within Katla's caldera shifted slightly toward the east, and a glacial flood washed out a nearby bridge.

Hekla may be the queen of Iceland's volcanoes, and Katla may be the most feared for its flood potential, but the title of most active volcano goes to Grímsvötn. Look at a satellite picture of Iceland and your eye will be immediately drawn to the big blob of white in the country's southeast. That's mighty Vatnajökull, the largest ice cap in Europe. Beneath it lies the subglacial lake known as Grímsvötn, created by a volcano with the same name that has erupted more than any other in Icelandic history. Chronicles from the sixteenth and seventeenth centuries talk about the *eldar*, or 'fires', that regularly come from this region.

Grímsvötn is a sort of natural geothermal plant. The equivalent of several thousand megawatts of continuous power create a meltwater lake atop the volcano, beneath the overlying ice. When enough water accumulates in the lake, the pressure forces an opening in the ice cap and the water pours out, draining most of the lake in a matter of days. This happens about once every five to ten years, although Grímsvötn's most recent eruption, in 2011, didn't cause any great flooding (perhaps because it had been drained relatively recently by another jökulhlaup). But the floods can be immensely powerful, leaving behind 'sand plains' that stretch for kilometre after kilometre all the way to the coast. The floods also knock out the sturdiest of Icelandic bridges: on the sand plains east of the town of Kirkjubæjarklaustur, you can stop to marvel at the twisted steel wreck of a bridge, tossed aside like a child's toy during a 1996 eruption.

Grímsvötn is but a single volcano, but it's the hub of a so-called fissure system, a network of long narrow cracks along which magma can rise to the surface. From the southwest corner of the Vatnajökull ice cap a couple of long linear features stretch out across the barren ground, as if a giant had reached down and scraped his fingernails across the landscape. These are the marks of some of the most powerful eruptions from the past millennium. Here, twice, lava has poured out in quantities greater than anywhere else in recent times. No wonder Icelanders call this area the Fire Districts.

The northernmost of these two great scrapes arrived in the year 934, in an event that takes its name from the mighty canyon known as Eldgjá, or 'fire fissure'. For six years it vomited more lava onto the landscape than any other eruption in the last millennium. Spread all the Eldgjá lava into a strip as wide as the Champs Elyseés, and it would run from Paris to Berlin as a wall 500 metres high. Mere decades after Iceland's first settlers had begun tilling the rich land, this pile-up of lava forced the Skaftá River to change its course. Many farmers fled, while others tried to clear their land as the deluge of ash fell onto it. They must have been choking on the eruption's fumes as they worked – Eldgjá also takes the title of the most polluting volcano of the past millennium. It pumped some 220 million tonnes of sulphur dioxide into the atmosphere, more than twenty times the amount released by Mount Pinatubo in the Philippines in 1991.

Almost no records of the terrible Eldgjá eruption survive, and local memories may have faded over time. But more than eight centuries after its lava cooled to vast featureless plains, Icelanders were to see a second set of violent eruptions in the Fire Districts. These were to come along a second set of giant fissures located just a few kilometres further south – on each side of a small mountain called Laki.

CHAPTER THREE

Supervolcanoes

The world's hotspots

NOWHERE IS THE VIOLENCE AND BEAUTY of our planet so apparent as in Yellowstone, America's first national park and home to what could be called the world's most dangerous volcano. Tourists flock to Yellowstone for the views: snow-capped peaks, lushly forested valleys, and teeming herds of bison and elk. Equally astounding is the park's vast array of geothermal features. 'Paint pots' of burbling mud spray red, white and other vibrant tints across a geological artist's palette. Bacterial mats the colour of autumn leaves stain the banks of the Yellowstone River. Steaming water percolates out of the ground and runs downhill, fashioning giant terraced and pillared structures out of minerals. And, of course, geysers like Old Faithful gush boiling water across the landscape. Yellowstone is home to a greater concentration of geysers than anywhere else in the world.

All this action arises from a hotspot of molten rock much like the mantle plume that lies beneath Iceland, though probably not as deep. The Yellowstone hotspot taps into heat about 200 kilometres below the surface, where rock melts and rises into a magma chamber just a few kilometres beneath the Wyoming

Geysers are plumes of water superheated by volcanic activity. A slumbering supervolcano beneath Yellowstone National Park, in the western United States, fuels an intense concentration of geysers there.

countryside. This giant reservoir is what heats groundwater above to create the many geysers, mud pots and hot springs.

Sixteen million years ago, the Yellowstone hotspot spewed out lava in what is today southeastern Oregon. Over time, as the crustal plate drifted westward above the hotspot, the plume punched out progressively younger tracks across the Snake River plain of southern Idaho. (The interstate highway that runs through Boise traverses these featureless plains of black lava. Idahoans have famously made the best of it by tilling the soil to grow potatoes.)

By the time the hotspot reached what is now Yellowstone, it let loose its wrath in three colossal eruptions. All three spread ash and debris – enough material to fill the Grand Canyon – over nearly the entire western half of the United States, and left behind colossal craters known as calderas, formed as magma beneath the ground drained out. The most recent eruption happened around 640,000 years ago and was about 2,500 times the size of the 1980 eruption of Mount St. Helens. The oldest of the three Yellowstone blowouts happened 2.1 million years ago and was 6,000 times bigger than Mount St. Helens.

That makes Yellowstone the quintessential example of a 'supervolcano', a word that has little technical merit but that is helpful for thinking about the relative scale of eruptions. BBC producers coined the term in 2000, to describe planet-altering eruptions that disgorge an immense volume of ash and other rock fragments – as a general rule, the volume of material that settles onto the ground is about two to three times the volume of the magma that fuelled the eruption.

In order to rank eruptions, volcanologists have developed the Volcanic Explosivity Index, or VEI, a scale that runs from 0 to 8. A VEI rating takes into account not only how much stuff a volcano blows out but also how high its eruptive plume rises. Each notch on the scale represents roughly a tenfold increase. An eruption with a VEI of 1, for instance, spits out more than 10,000 cubic metres of material and has a plume between 100 and 1,000 metres high. (Italy's Stromboli does this almost constantly.) A VEI 2 eruption will blow out more than 1 million cubic metres of stuff, with a plume between 1 and 5 kilometres high. This kind of eruption might go off once a week somewhere around the globe. Going on up the scale, a VEI 3 eruption happens perhaps annually.

After that, things get really interesting. 'Cataclysmic' eruptions with a VEI of 4, such as the 2010 eruption of Eyjafjallajökull or the 1783–84 eruption of Laki, send a plume between 10 and 25 kilometres high. The archetypal VEI 5 eruption (aka 'paroxysmal') is the May 1980 eruption of Mount St. Helens, which ripped the top off the mountain and sent the largest recorded landslide roaring down the valley below. A VEI of 6 ('colossal') is something that happens only once every century – like Mount Pinatubo, which blew up in the Philippines in 1991, sending ash more than 30 kilometres high to encircle and cool the globe for several years. As if that weren't enough, a VEI 7 eruption gets dubbed 'super-colossal' for spewing more than 100 cubic kilometres of material. The last time this happened was in Tambora, Indonesia, in 1815.

And then there are the VEI 8 eruptions, which start to edge off the scale of possible description. 'Mega-colossal' the ranking scale calls them. These furious explosions send out more than 1,000 cubic kilometres of debris, with plumes reaching more than 50 kilometres high. That's almost to the edge of space.

Yellowstone's eruption 2.1 million years ago ranked a VEI 8. But that's far beyond the reach of human history. To understand what a VEI 8 eruption can do in more recent times, you have to look to Toba, Indonesia. The most recent VEI 8 blast happened there, 74,000 years ago, and is almost beyond believability.

Toba was destined to blow from the start. The volcano nestles on the northern part of the island of Sumatra, where the Indo-Australian tectonic plate dives beneath the Sunda plate. The collision fuels the chain of fire mountains that make Indonesia the most volcanically active place in the world.

In the past million years or so, Toba has erupted four times, culminating in the one 74,000 years ago. Numbers such as 'VEI 8' don't really do justice to this sort of devastation. When volcanologists think about the worst-case eruption, the kind of colossal blast that changes the planet forever, Toba is what they think of.

The eruption blasted an ash cloud across most of Southeast Asia. Landscapes were buried. Rivers choked and stopped flowing. The first scientists to identify Toba's ash thought the

The eruption of Indonesia's Toba volcano, 74,000 years ago, left behind a 100-kilometre-long lake – clearly visible in this space shot.

eruption must have covered some 5 million square kilometres, itself an astounding number. Other geologists later found the Toba signature even further away, in seafloor cores taken from the South China Sea and the Indian Ocean south of the equator. That geographical spread roughly doubles the amount of ash that Toba must have disgorged. In all, the volcano buried 1 per cent of the planet's surface at least 10 centimetres deep in debris.

Rocks falling out of the sky would have levelled trees and killed animals on the ground. The skies would have darkened noticeably, to be somewhere between an ordinary overcast day and something almost as dim as a moonlit night. The combination of the lava, the rocks and the gases belching out of Toba would have eradicated all life around the mountain. The neighbouring island of Mentawai, 350 kilometres away, somehow escaped the worst of the damage because it remains home to monkeys and squirrels that would have become extinct had Toba's debris fallen heavily on the island's rainforests.

Further afield, the sulphur from the volcano would have quickly turned into small particles that reflected the sun's incoming rays back into space, shading and cooling the planet beneath. In fact, Toba has become something of an archetype for scientists who are trying to understand how volcanic eruptions can affect global climate. The first research into its climatic effects took place in the 1970s, around the time researchers were beginning to assess the atmospheric impacts of a possible nuclear war between the United States and the Soviet Union. It turns out that debris from a nuclear bomb would spread around the globe much as an ash cloud does, cooling and triggering a 'nuclear winter' that could freeze people and crops worldwide. Similarly, geologists proposed that an eruption such as Toba could set off a 'volcanic winter' that would devastate the planet for years.

To figure out how temperatures changed, some researchers look for environmental records such as fragments of ancient plants, while others use state-of-the-art climate models to simulate how temperatures may have fallen as the volcanic pall spread around the globe. General consensus today holds that Toba caused global temperatures to plummet by as much as 10 degrees Celsius for a year or more. The climate would have stayed dramatically colder for a decade or two, followed by generally cooler temperatures and lower precipitation for several more centuries.

Most controversially, the eruption may have even affected human evolution. Toba went off at around the time that modern humans, *Homo sapiens*, were moving out of their evolutionary birthplace of Africa and into Asia. It's hard not to wonder what those newcomers must have thought when they looked up and saw a towering ash cloud moving toward them. Scientists have long known that early humans went through population 'bottlenecks', in which many died out and only a certain number emerged to continue passing their genes from generation to generation. In 1998, Stanley Ambrose of the University of Illinois linked one of these population bottlenecks to the Toba eruption. For early humans desperately looking for things to eat, the big chill would have been just too much to handle, Ambrose proposed. 'Toba's volcanic winter could have decimated most modern human populations,' he wrote in the *Journal of Human Evolution*.

Ambrose's proposal drove straight to the heart of some of the biggest questions in science. Where did we come from, and how did we get here? How did our environment shape us, and how resilient were we in the face of disaster? Do we have any control over our destiny, or are we constantly subject to the whims of nature looking to wipe us out? Naturally, other researchers started looking for answers – mostly in India, which was smack in the path of Toba's ashfall.

In southern India's Jurreru Valley, a team of British and Indian scientists has found stone tools and other artefacts that they argue were made by *Homo sapiens* and date to both before and after the Toba eruption. If so, the discovery suggests that early humans were able to weather the eruption without too much trouble. Other researchers argue that perhaps Toba forced people to retreat to nearby refuges until conditions improved. Further north, in north-central India, environmental records show how the Toba eruption caused vegetation to shift from forests and woodlands to mainly open grasslands for at least a couple of centuries – a change that must have forced people to find something new to eat.

Toba's eruption was crucial to the history of life on Earth. And it may one day make another intervention. The blast left behind a caldera 100 kilometres long and 30 kilometres wide, which is now the world's largest volcanic lake. Modern measurements show that the bottom of Lake Toba is rising, perhaps because of magma filling the chamber beneath the legendary volcano. It is likely to be a few more hundred thousand years before Toba clears its throat again.

It won't be nearly as long before another feared volcano goes off.

Today the Greek island of Santorini (or Thera) is a tourist's paradise, with azure roofs and whitewashed buildings gleaming above the Aegean Sea. But Santorini's tranquillity belies its fiery underbelly. It is one of a strand of volcanoes created as Africa's tectonic plate slides beneath Eurasia, causing magma to rise from below.

Santorini most famously blew its top around 1600 B.C.E., during the Bronze Age. The eruption was so violent that it shattered the volcano and left behind an expansive caldera, like that at Yellowstone. But here, instead of being filled with aspen and bison, the caldera is filled with the waters of the Aegean. The islands known collectively as Santorini outline part of the caldera's rim. The steep cliffs on which homes are perched underscore the eruption's violence: the cliffs are composed of layer upon layer of volcanic ash that piled up hundreds of metres thick and then consolidated into rock.

Debris from this blast, usually called the Thera eruption, can be found all over the eastern Mediterranean. The eruption, rated a VEI of 6 or greater, would have sent tsunamis in all directions, drowning sailors and coastal villages. Some scholars have even linked the explosion to the decline of the mighty Minoan civilization, which thrived on and around the island of Crete. Remains of Minoan pots have been found buried beneath ash from Santorini, which does seem to support the idea that the Bronze Age eruption hastened the decline of the Minoans. But archaeologists don't have a well-defined timetable for when particular events happened in this region, so it is hard to know for sure.

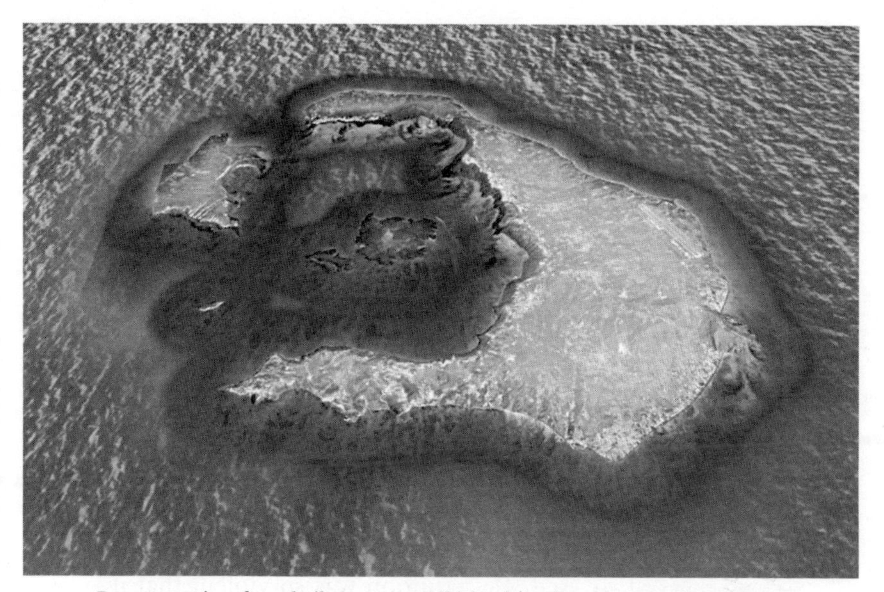

Recent eruptions have built up two small islands in the massive collapsed crater left behind by the Bronze Age eruption of Santorini, Greece.

Stories of the eruption must surely have survived in oral histories, however, and it's highly plausible that, some 1,300 years later, Plato was thinking of these tales of Thera when he wrote about Atlantis – an island that sank beneath the sea, accompanied by flames and a shuddering of the Earth. In his *Timaeus*, Plato describes the place:

> *Now in this island of Atlantis there existed a confederation of kings, of great and marvellous power, which held sway over all the island, and over many other islands also and parts of the continent . . . But at a later time there occurred portentous earthquakes and floods, and one grievous day and night befell them, when the whole body of your warriors was swallowed up by the earth, and the island of Atlantis in like manner was swallowed up by the sea and vanished.*

Of course, no one can be certain whether Plato was relaying a tale that had been passed down over generations and was based in fact, or simply speculating on a lost world. But a disaster such as the Thera eruption would not easily be forgotten.

Santorini has more to come. The volcano has been occasionally active since its Bronze Age blowout, and over the past few centuries small eruptions have built up two little islands in the caldera's centre, named Palaea Kameni (meaning 'old burnt island') and Nea Kameni ('new burnt island'). Greek authorities are worried enough about Santorini to have helped an international team of researchers to set up a monitoring system across the caldera. Global positioning system stations dotted around Santorini's rim receive signals from satellites, which measure tiny changes in the stations' elevation. Rises of just a few millimetres a year could indicate that a magma chamber beneath Santorini is beginning to inflate, like a person drawing in a big breath. This kind of ground deformation is often seen before a volcano erupts.

The stations captured such an inflation event starting in January 2011, with the ground rising 140 millimetres in 12 months. But volcanoes can uplift land without actually erupting; in the 1980s, the ground near Pozzuoli, Italy, rose nearly two metres, but the Campi Flegrei caldera below it did not erupt. Santorini's inflation slowed to a near-stop in 2012, and it's not clear whether the volcano plans to erupt anytime soon. Even if it does, geologists say, it won't be anything like the scale of the Minoan eruption. It will probably be something like the eruptions of the 1950s: small and burbling and possibly building up new small islands. Santorini is not dead yet.

Neither is one of the world's most notorious volcanoes: Vesuvius, the destroyer of Pompeii. This iconic mountain, which glowers over the city of Naples, belongs to one of Europe's most active volcanic fields. Within a small geographic span lie the fiery fountains of Etna, the fuming peak of Stromboli, and the bubbling mud pots of the Phlegraean Fields – home to a supervolcano that exploded some 39,000 years ago.

Even among all this competition, Vesuvius is the granddaddy of Italian volcanoes, and its eruption in 79 C.E. is the founding narrative for all disaster stories that followed. The catastrophe was captured in words by Pliny the Younger, who described the horror of the event and the death of his uncle, the polymath Pliny the Elder. Centuries later, archaeologists unearthed even more eloquent testimony in the form of Pompeii and Herculaneum, towns eerily entombed in layers of volcanic ash.

Vesuvius has been active for millennia. Gazing at the mountain, the ancient geographer Strabo recognized this violent history in the rocks: 'One might infer that in earlier times this district . . . had craters of fire,' he wrote. The historian Plutarch later wrote of the sky around Vesuvius as being 'on fire', and the poet Silius Italicus described the mountain 'hurling flames worthy of Etna from her cliffs'. But in the years leading up to 79 C.E., nobody was paying much attention. Thousands of people lived obliviously on the gentle slopes around the mountain, where vineyards thrived in the fertile soil. Life seemed good.

On 23 August of that year, residents in Rome celebrated the annual Volcanalia festival, in an effort to appease the god of flaming mountains (or, perhaps, just to throw a good party). Apparently Vulcan wasn't paying attention, because on the afternoon of 24 August, Vesuvius suddenly blew its top. Ash shot into the air, spreading quickly into a threatening cloud. In Pompeii, ten kilometres downwind, stones and ash began raining from the sky. People fled in terror as ash piled on the roofs of buildings, collapsing them. In less than a day, debris buried parts of the town as much as 1.4 metres deep.

Just after midnight, the pyroclastic flows began. These surges of hot rock and ash avalanched down the mountain's western slopes. Herculaneum, just a few kilometres away, had no chance: the street grid funnelled the fiery flows directly into the heart of the town. People, animals, buildings – everything was incinerated. Unlike Pompeii, those in Herculaneum had practically no time to escape.

Across the bay to the west, two of history's most famous scribes watched the scene with mingled fascination and horror. Pliny the Elder had been staying at his summer villa with his sister and her son, the 17-year-old Pliny the Younger. When Vesuvius exploded, the elder statesman apparently couldn't believe his luck at having an opportunity to witness such a momentous eruption. He climbed to the highest point he could find and watched the ash cloud develop. Later his nephew recalled how they saw the plume grow into a shape 'like an umbrella pine, for it rose to a great height on a sort of trunk and then split off into branches, I imagine because it was thrust upwards by the first blast and then left unsupported as the pressure subsided.' This classic, umbrella-like shape forms when a volcanic ash plume hits winds going sideways in the upper atmosphere and spreads out. We now use the adjective 'plinian' for this type of ash cloud, after the man who first described it.

Pliny the Elder then decided to take a boat and look more closely at what was happening. His nephew wisely declined. Within hours the elder Pliny was dead, having collapsed on the beach where he had gone to check out the eruption. Two letters from the young man, written to a friend three decades later, describe the horror of the hours that followed:

> *We had scarcely sat down to rest when darkness fell, not the dark of a moonless or cloudy night, but as if the lamp had been put out in a closed room. You could hear the shrieks of women, the wailing of infants, and the shouting of men; some were calling their parents, others their children or their wives, trying to recognize them by their voices. People bewailed their own fate or that of their relatives, and there were some who prayed for death in their terror of dying. Many besought the aid of the gods, but still more imagined there were no gods left, and that the universe was plunged into eternal darkness for evermore ... I could boast that not a groan or cry of fear escaped me in these perils, had I not*

derived some poor consolation in my mortal lot from the belief that the whole world was dying with me and I with it.

Yet even after such agonies, the story of Vesuvius in 79 CE faded into local memory. Its tale did not come to light until more than a millennium and a half later, when people began to realize that entire Roman towns lay buried beneath their feet. In 1592, a canal excavation near Naples uncovered what would be the first artefacts from Pompeii. Just over a century later, digs began in earnest when a prince bought some land in

This Roman fresco celebrating Bacchus, the god of wine, was unearthed at Pompeii, Italy. It shows Vesuvius before the volcano blew its top in 79 C.E.

the area and, in the process of sinking a well, found a subterranean town. By 1738 much of Pompeii was unearthed. The finds were a remarkable glimpse into Roman life: mosaics, paintings and foods preserved as they were when Vesuvius entombed them. People were incinerated instantly as the ash congealed around them, forming moulds of their bodies at the moment of death.

Soon, Vesuvius became the laboratory in which leading ideas about volcanoes were forged, modified or rejected. A British government representative, Sir William Hamilton, lived in Naples from 1764 until 1800 and climbed the mountain time and again, bringing along any curious visitors he could entice to join him. He sent long reports of Vesuvius's activity to London's Royal Society, and even shipped back a couple of light-and-sound devices that displayed the mountain pouring forth streams of lava to the accompaniment of booming thuds. More than anyone before him, Hamilton brought Vesuvius into public consciousness beyond Italy.

Since 79 CE, Vesuvius has erupted more than fifty times. It has been silent since a small eruption in 1944, but those who know Vesuvius say it is only a matter of time. Volcanologists recently identified a zone, about eight to ten kilometres below the mountain, where seismic waves travel more slowly than usual. That zone could be a magma chamber slumbering away – for now. Chemical studies of erupted lava suggest that the stuff in the underground reservoir could be capable of immediately creating a massive eruption. Looming over a city of more than three million people, the volcano defined by Pliny may soon show the world what it did two millennia ago.

Our narrative now skips forward to a year much closer to modern times. In 1815, just a few decades after Laki cloaked much of Europe in a toxic fog, another deadly volcano sent its gaseous tendrils around the world. This time the culprit lay half a world away, in the volcanically restive islands of

Indonesia. The VEI 7 eruption of Tambora is the greatest known to history. More than 70,000 people perished, the most known from any single eruption, many of whom sickened or starved as ash buried their rice fields. Further afield, millions shivered as the volcano spread its aerosols worldwide, chilling the planet and causing the famous 'year without a summer' of 1816.

Before Tambora blew, the central Indonesian island of Sumbawa was a tranquil and productive place. 'Nature had poured its bountiful blessings on this island,' wrote a pair of visitors in 1824, which 'no matter how mountainous, is the proud possessor of the most extensive of plains and the loveliest of verdant valleys. Rice, beans and maize were plentiful, the forests provided wax and excellent timber ... Coffee, pepper and more especially cotton were grown.' Locals collected birds' nests and salt to sell, and the island was famous for the quality of its horses.

Looming over all this bounty was the mighty peak of Tambora. At some 4,000 metres high, it was one of the tallest mountains in all of Indonesia and a landmark for sailors to navigate by. And there were plenty of sailors to see it: in the spring of 1815, Indonesia was still the Dutch East Indies, the 'spice islands' from which Europeans shipped back pepper, nutmeg and other fragrant commodities. Trade was thriving, and the waters around Sumbawa bustled with trading vessels, fishermen and pirates.

In the midst of all this activity, Britain had managed to wrest control from the Dutch and occupy the island of Java in 1811. In the capital, Batavia (today's Jakarta), Thomas Stamford Raffles served as lieutenant governor. Raffles would be there to observe Tambora's blast and collect accounts of the disaster from around the archipelago, building the best-yet documentation of the effects of a massive eruption.

Tambora had probably lain quiet for more than a millennium before it began half-heartedly spitting out ash in 1812. Then, on the evening of 5 April 1815, the mountain awoke violently. Ash and smoke began to rise, and explosions shook the ground. Thousands of kilometres to the west, in Batavia, Raffles thought a ship might be putting out distress calls and sent his troops to check. Further east, on Java, soldiers thought they heard cannon fire and marched out to look for attackers.

Five days later, in a colossal eruption unlike any ever witnessed, Tambora hurled smoke and ash even higher, perhaps as high as 40 kilometres. A local chief later described the scene to one of Raffles's officers:

> *Three distinct columns of flame burst forth . . . and after ascending separately to a very great height, their tops united in the air in a troubled and confused manner . . . In a short time, the whole mountain ... appeared like a body of liquid fire, extending itself in every direction. The fire and columns of flame continued to rage with unabated fury, until the darkness caused by the quantity of falling matter obscured it.*

Floods of superheated gas and ash surged down the mountain's sides, wiping out villages at its base. The flows plunged into the ocean, reacting with seawater and sending mighty ash plumes hurtling skyward. For three days, day turned to night. Wading through the ash-choked darkness, people could not see their hands held before their eyes. When the daylight returned, they were shocked to see that Tambora had lost more than 1,000 metres from its summit.

Death and destruction continued to rain down. Ash buried buildings, which subsequently collapsed under the weight, killing people sheltered there. A tsunami swept across northern Sumbawa, drowning villages. And the horror didn't end when Tambora quietened down. Thick blankets of ash covered the rice fields, smothering crops. The once-bountiful island began to starve. Some people sold their children for bags of rice imported from other islands. Water contaminated with ash gave anyone who drank it rampant diarrhoea. Corpses began to pile up by the sides of roads.

The 1815 eruption of Tambora drastically reduced one of Indonesia's tallest peaks (photographed here by astronauts aboard the International Space Station).

With little way to make sense of such catastrophe, the locals turned to the spiritual for explanations. Soon a folk tale sprang up, in which the ruler of Tambora's kingdom had slaughtered an innocent Muslim pilgrim. The volcano erupted, the story goes, as divine retribution for this violation. A local poem describes the shame:

Its noise reverberated loudly
Torrents of water mixed with ash descended
Children and mothers screamed and cried
Believing the world had turned to ash
The cause was said to be the wrath of God Almighty
At the deed of the King of Tambora
In murdering a worthy pilgrim, spilling his blood
Rashlessly and thoughtlessly.

The disaster didn't end in Indonesia. The sulphur emitted by Tambora would go on to combine with water and produce acidic particles that spread around the globe, altering weather worldwide. By the following year temperatures had plummeted in western Europe and northeastern North America, earning it the nickname of 'eighteen hundred and froze to death'. In early June, a cold front swept across the northeastern United States, such that snow fell in Albany, New York, and frost killed most of the apple fruit that had just finished blossoming. Unbelievably, hard frosts came again in July and August. The major crops of corn and hay failed, leaving little fodder for livestock to eat the following winter. Disheartened, many farmers struck out for what they hoped was better weather in the west, hastening the American migration westward.

Europe fared little better. That summer turned out to be the coldest in Britain since 1750. Heavy rains fell across western Europe, drowning crops in the field. In July 1816 in Alsace, a farmer wrote: 'The rainy weather continues. The hay has not been made anywhere. The grass is rotting on the meadows, all mountains are full of water. There is nothing but misery everywhere.'

Perhaps the only bright spot in that dismal and depressing summer of 1816 was its gift to literature, though it was a dark offering. The poet Lord Byron – escaping a disastrous marriage, accumulating debts and allegations of unseemly behaviour in England – had settled at a country villa near Lake Geneva. The soft summer weather he expected, however, had fallen into a consistently cold and rainy trend. Soon thereafter, he met up with the poet Percy Bysshe Shelley and his wife Mary, and the couple moved into Byron's manor. Although they had hoped the weather would break, it remained too inclement for outdoor activities. Trapped and bored inside their rented villa, the group decided that, to pass the time, each of them would write a ghost story. Thus was born Mary Shelley's greatest creation, *Frankenstein.*

Not to be outdone, Lord Byron generated his own tale of gloom: the poem 'Darkness', which begins with these glum lines:

I had a dream, which was not all a dream.
The bright sun was extinguish'd, and the stars
Did wander darkling in the eternal space,
Rayless, and pathless, and the icy earth
Swung blind and blackening in the moonless air;
Morn came and went – and came, and brought no day,
And men forgot their passions in the dread
Of this their desolation.

Byron goes on to depict ships rotting at sea, famine preying upon entrails, and dogs turning upon and eating their masters. Suffice it to say, 'Darkness' is a downer. But it accurately reflects the gloom that enveloped much of the world in 1816 – perhaps because of political unrest following the Napoleonic wars, but also in part because of the obscure volcano that had erupted in Indonesia.

Visual arts may also have benefited from Tambora. The volcanic aerosols scattered the sunlight, dramatically reddening sunsets for years after the eruption. It has been speculated that some of the fantastically coloured skies painted by J.M.W. Turner were in part the creation of Tambora.

But as agriculture failed, famine soon arrived in Europe. Families became refugees, begging for food or scavenging in the fields for rotted crops. Prices for staples such as bread rose dramatically, especially in the rural areas. In England, hungry farmers rioted and torched barns. In Switzerland, government officials offered to teach people to distinguish poisonous from non-poisonous plants, so that they could forage on the hillsides safely.

Weakened by lack of food, people became more prone to disease. Typhus ravaged the British Isles, and a doctor in Ireland wrote: 'I consider the predisposing causes of the present Epidemic to have been the great and universal distress occasioned among the poorer classes, by the scarcity which followed the bad harvest of 1816, together with the depressed state of trade and manufactures of all kinds.' Around the same time, the world's first true cholera epidemic began in Bengal. Some scientists have linked this, too, to the general vulnerability of people following the explosion of Tambora.

Missing in all this misery was an explanation of its cause. Europeans had read Indonesian reports of Tambora's massive eruption, but few thought to link the volcano to weather changes halfway around the world. Struggling to understand what had happened, researchers generated a long list of possible causes for the year without a summer. Some scientists thought the cold in western Europe must come from sea ice that had discharged from the Arctic and drifted south in the North Atlantic. Sunspots were another popular theory: fewer sunspot numbers led to colder weather, many speculated. Others blamed the recent introduction of lightning rods, which were thought to discharge the electricity that triggers cloud formation, hence allowing far more rain than normal to fall. Some people, of course, blamed God.

Not until the early twentieth century did scientists finally link the year without a summer to Tambora. In 1912 the Katmai volcano in Alaska blew up, in what would be the largest eruption of the century. Meteorologist W. J. Humphreys, noting the effect of the volcanic dust on surface temperatures, linked Tambora to the extreme cold of 1816 and claimed that volcanoes could be responsible for many past climate changes. 'Through it, at least in part, the world is yet to know many another climatic change in an irregular but well-nigh endless series,' he wrote. 'Usually slight though always important, but occasionally it may be, as in the past, both profound and disastrous.'

Profound and disastrous indeed was the next great eruption, Indonesia's Krakatau in 1883. Krakatau once rose majestically in the heavily travelled Sunda strait between Java and

Sumatra. It lay 1,400 kilometres west of Tambora, but with nearly seven decades having passed since that devastation, many Indonesians had forgotten what their mountains could do. Krakatau changed all that. It would annihilate itself and some of the loveliest areas of Indonesia, sending huge tsunamis racing across its vulnerable environs. Within a day, 36,000 people would be dead.

Krakatau had been quiet for millennia before it began spouting a few ash plumes throughout the summer of 1883. Then, with little warning, it erupted with a vengeance on 26 August. The captain of a German vessel passing through the Sunda strait spotted it first, a white cloud rising from the mountain's summit. The cloud soon reached high into the atmosphere and spread out into the characteristic umbrella shape of a plinian eruption. That afternoon, ash and pumice began to pour from the sky. Then at 3.34 p.m. the first major explosion shook the mountain; three minutes later the first tsunami rushed into the main canal of Batavia, 150 kilometres away. Explosions fired off again and again, reverberating across the ocean and surrounding islands. Larger pieces of pumice began to fall, some still warm to the touch.

In the darkest hours the eruption began sending pyroclastic flows directly into the ocean. The captain of a passing ship described the eruption: 'Chains of fire appeared to ascend and descend between it and the sky, while on the southwest end there seemed to be a continual roll of balls of white fire.' Daybreak was barely visible through the black clouds that obscured the sun. Explosions thundered on and on until Krakatau obliterated itself in a final mighty roar, at precisely 10.02 a.m. The detonation was so loud that people heard it in Singapore, Ceylon, and on Rodrigues Island, more than 4,700 kilometres away.

Some of the pyroclastic flows were so fast and furious that they travelled on top of the sea, hot enough to burn victims as far as 80 kilometres from the eruption. A survivor in Sumatra, some 40 kilometres north of Krakatau, recalled the fear:

Krakatau's 1883 eruption, in Indonesia, sent sound waves reverberating around the globe seven times.

'Suddenly it became pitch dark. The last thing I saw was the ash being pushed up through the cracks in the floorboards, like a fountain . . . I felt a heavy pressure throwing me to the ground. Then it seemed as if all the air was being sucked away and I could not breathe.' The writer, who was the wife of a local government official, escaped alive but only barely. Her skin was badly burned, and only when she tried to nurse her infant son did she realise he was dead.

Nearly all of Krakatau's victims perished not by fire but by water. The volcano's violent shuddering kicked up tsunami after tsunami, which raced across the ocean and swept across the neighbouring islands. Wave heights reached 40 metres. Boats were picked up and tossed kilometres inland, and entire villages were engulfed. On the island of Sebesi, 15 kilometres north of Krakatau, every single one of its 3,000 residents was drowned.

Wave after wave obliterated coastal villages that were little more than clusters of mud and stick huts. No one knew the tsunamis were coming until they arrived, running up suddenly to great heights as they shoaled upon the beaches. Because of the flatness of the Indonesian coasts, those away from the beach weren't safe either. One farmer who was working in his rice fields, around eight kilometres inland in Java, described 'a great black thing ... very high and very strong' coming towards him. Everyone ran for higher ground. His survivor's tale could not be more grim: 'There was a general rush to climb up in one particular place,' he wrote:

> *This caused a great block, and many of them got wedged together and could not move. Then they struggled and fought, screaming and crying out all the time. Those below tried to make those above them move on again by biting their heels. A great struggle took place for a few moments, but ... one after another they were washed down and carried far away by the rushing waters. You can see the marks on the hill side where the fight for life took place.*

Corpses piled up in the water, so thick you could walk across them. Some became lodged in rafts of floating pumice. Eleven months after the eruption, African schoolboys playing along the beach in Zanzibar found pumice stones strewn along the beach; human skulls and bones were piled up at the high-water mark.

Krakatau's 1883 eruption was the loudest explosion ever heard. It rang the planet like a bell, the sound waves encircling and reverberating around the globe seven times. A pressure gauge at the gasworks in Batavia recorded an atmospheric pressure spike of more than two and a half inches of mercury, blowing right off its scale.

In the end, Krakatau erupted for 100 days before sinking beneath the ocean. A telegram sent from Batavia read simply: 'Where once Mount Krakatau stood the sea now plays.' Since then a small island, known as Anak Krakatau or 'child of Krakatau', has risen from the waters in its place. It smokes and belches occasionally, but shows no signs of other activity.

For months Krakatau's aerosols spread around the globe, reflecting sunlight and creating lurid colours in the sky, just as Tambora had done. The moon appeared blue at times, as did the sun. Sunsets were a spectacular marbling of red, gold, orange and other hues. In New York and Connecticut, people called the fire department because they thought the red glow on the horizon was from a conflagration. The fantastically coloured cloud bands in Edvard Munch's famous painting *The Scream* might have been inspired by Krakatau-tinted skies that the artist had seen in Norway. Even Alfred Tennyson took a stab at describing the spectacle in his poem 'St. Telemachus': 'Had the fierce ashes of some fiery peak/Been hurl'd so high they ranged about the globe?'

Just a century after Laki erupted, Krakatau showed just how far global science and communication had come. By 1883 the first seismometer arrays had been installed in Indonesia and elsewhere. Tide gauges recorded the phenomenal tsunamis. Most significantly, submarine telegraph cables linked Indonesia to the rest of the world, allowing messages to cross the world in hours. Stories from Batavia streamed back to Europe, where newspaper readers hung on every word of the fabled disaster.

In his book *Krakatoa*, Simon Winchester makes the argument that Krakatau was 'the event that presaged all the debates that continue to this day: about global warming, greenhouse gases, acid rain, ecological interdependence. Few in Victorian times had begun to think truly globally . . . Krakatoa, however, began to change all that.'

But the seeds for this change had been planted a century earlier, in the Sída district of Iceland. There, Jón Steingrímsson was struggling to make sense of the unimaginable disaster that had just descended upon him and his parish.

CHAPTER FOUR

Fire, Famine and Death

The poisoning of Iceland

SUNDAY MORNING, 20 JULY 1783, dawned on the village of Klaustur devoid of light and hope. By now, everyone knew that the snout of Laki's lava was just around the bend in the Skaftá River channel, not three kilometres to the west as the crow flies, and drawing ever nearer. The sky was heavily overcast, the air sulphurous and acrid, charged with lightning and reverberating with unnerving thunder. For forty-two days and nights the fires of the earth had assailed the villagers, stifling them with fumes and humbling them with the awesome destructiveness of molten rock. It was clear to Jón that God's retribution was far from over. Perhaps the Lord was testing him, his faith and his ability to inspire his dispirited flock.

The villagers had fair reason to feel bereft. Some, just days before, had watched as fires overran their villages, homes and farms. And now the lava had nowhere else to go except straight down the river gorge toward Klaustur. Those fit enough would be forced to flee southward toward the sea, where the lava was sure to follow, or scramble up the steep cliffs to the north and make their way along precipitous paths worn by grazing sheep. The old and the infirm would have to fend for themselves.

And so on Sunday the people gathered for services, perhaps for the last time, in their only remaining refuge – the church. The journey must have been grim. Ash choked the air and cut off the sun. As the villagers passed through the gloom, they could glimpse only the low stone wall and arched gate at first, then the chapel's steeple and cross. Inside the church, the only light came from a succession of lightning flashes, followed by claps of thunder so loud the steeple bells chimed in response.

Jón couldn't help thinking about two other churches in the region that had already been consumed by lava. It looked entirely possible that today his would meet the same fate. But there wasn't much time to dwell on such gloomy thoughts. Once everyone was assembled in their usual rows, Jón took his place at the pulpit. What he was about to do would become legendary in Iceland and earn him the nickname Eldprestur or 'Fire Priest'. The words he was about to speak would, to this day, be known as the Eldmessa or 'Fire Mass'.

Oddly, Jón – a man so adept at chronicling local history – left behind only a few sparse sentences describing his Fire Mass. But here's what we know happened. He began by urging everyone to pray to God 'with proper meekness', and to ask that in His mercy He should not be so quick to destroy them or the church. Great as the calamity was, 'just as great was His almighty strength in our weakness', Jón told his parishioners. Tumult raged outside as he spoke, but those inside the church appeared calm and resigned to their fate:

> *Each and every person was without fear, asking His mercy and submitting to His will. I have no reason to believe otherwise than that every man was prepared to die there, if this would have pleased Him, and would not have left even if things had become worse, because now it was impossible to see where there was a safe place.*

Jón's impassioned entreaty drew out the service a little longer than usual. Under the circumstances, he felt, no length of

This modern altar from the Prestbakki church commemorates Jón's famous Fire Mass, which he preached as lava threatened to destroy his village.

time spent talking to God could be too long. The churchgoers apparently agreed. Even though they all felt the lava would soon be upon them, no one tried to flee.

Finally the service ended. The villagers ventured outside, wondering if their homes had been swallowed by lava. But they could see no new devastation. No molten rock, no fires, no destruction had come upon them as Jón had preached. Puzzled, a group of men hiked up the river channel to see how much closer the lava had come. To their astonishment, they saw that it had advanced not a metre during the service. 'During the time which had elapsed,' Jón later wrote, 'it had collected and piled up in the same place, layer upon layer, in a downward-sloping channel some 70 fathoms wide and 20 deep, and will rest there in plain sight until the end of the world, unless transformed once again.' What seemed like an inexorable advance had, inexplicably, halted.

Today, the place where the lava stopped is still visible, a few kilometres upstream from the spot where the Klaustur chapel stood. It is a low, weathered wall of black rock in the middle of the Skaftá River channel. It seems a comfortable enough distance from town, but only because we know the outcome of the story. For those suffering that day in 1783, the halting of the lava must have seemed a wondrous deliverance. Perhaps God had intervened on their behalf after all.

It was, indeed, a miracle of sorts – but not necessarily one of divine intervention. As we have seen, Jón was a keen observer of the natural world. He had been watching the lava advance and noting how it flowed along the Skaftá channel. He also knew that heavy rains had swollen two tributaries of the Skaftá, which fed into the river gorge just upstream of Klaustur. These two tributaries turned out to be crucial in explaining what happened during the Fire Mass. As Jón describes in his journal:

> *The rivers Holtsá and Fjathará poured over the dams which the new lava had made them, and with great torrents and splashing smothered the fire, which was churning and rumbling in the channel, then poured forwards and off the front of the aforementioned pile, streaming and splashing. There was so much water that horses could not cross the river at all by the cloister all that day.*

This powerful flood inundated the lava, cooling and solidifying it and halting its advance. The miracle was not that the lava halted; the miracle was that the rains and resulting deluge came when they did.

Did Jón know that the lava flow would stop? Did he co-opt a phenomenon of nature to pull off a seeming miracle in front of his congregation? We'll never know for sure. Jón credited God with saving his village in July 1783, and to many that explanation would suffice. Others surely didn't care what had stopped the lava, as long as it did stop. In any case, the Fire Mass turned Jón into an eighteenth-century celebrity, the 'fire priest' whose faith could stop a volcano.

For Jón, the halting of the lava was an almost indescribable relief:

> *We left the church more cheerful than I can describe and thanked God for the very visible protection and deliverance that He had granted to us and His house. Yes, and may everyone who sees this almighty work and hears of it spoken whether alive or yet unborn, praise and proclaim His worthy name. From this day onwards the fire did no major damage to my parish in this way.*

Yet the consequences of the Laki fires would damage his parish in all sorts of awful ways.

The days following the Fire Mass turned out to be a brief interlude in a lengthy drama. Five major eruption episodes had already occurred along a line southwest of Mount Laki, each opening up new surface fissures. Much of the lava travelled along the Skaftá gorge that had once been as much as 200 metres deep and 60 metres wide, but was now filled to the brink with molten rock. A second fork of lava flattened itself out on the floor of the Varmárdalur valley and eventually merged with the main Skaftá flow.

The combined rivers of lava raced down the gorge – sometimes covering six kilometres in a day – and emerged onto broad cultivated plains at its mouth. Unable to veer east towards Klaustur, because of the quenched flows in that direction, the lava devastated the southern lowlands instead. Layer after layer of black, barren rock buried what had once been verdant fields full of sheep, cattle and farm buildings. By early September that flow ceased.

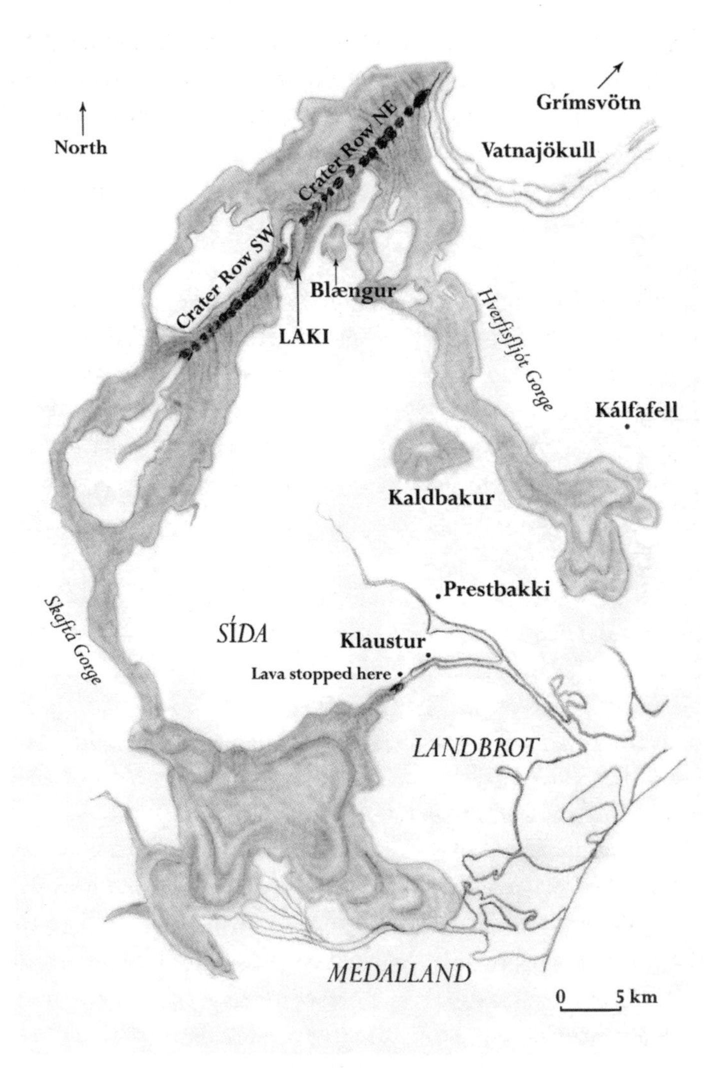

Between June 1783 and February 1784, two great arms of lava flooded down from Laki to embrace each side of the town of Klaustur.

But just days after the Fire Mass, unbeknownst to Jón or anyone else, the rift was beginning to rend the landscape north and east of Mount Laki. The great seam that had opened up in the earth on 8 June was now lengthening. This shift would change the character of the eruption and create new bedlam for residents in Klaustur and in the lowlands to the east. (The modern volcanologist Thorvaldur Thordarson has unravelled these steps of the eruption in great detail, discovering that at least ten separate episodes made up what Jón and his parishioners saw as two phases of the eruption.)

The herald of this new phase, Jón wrote, was an intense fiery glow observed in the sky on 24 and 25 July, followed on the 28th by rain and thunder that brought fine ash and a strong foul smell. On 29 July, Jón spotted 'a frightening cloud' rising into the sky, accompanied by rumbling and cracking sounds. A second black cloud appeared and blocked the sun, bringing absolute darkness and a shower of fine ashes. The next day, although the weather was mild, 'thuds, cracking, and thundering' could be heard without pause, coming seemingly from all directions.

On 31 July, it became obvious that Laki's lava was about to vapourize another major river. To the north and east of Klaustur, ominous clouds of steam could be seen moving along the gorge of the Hverfisfljót River. In some of its channels, water boiled as if in a cauldron. The destruction that had been visited on parishes to the west was clearly about to be visited on those to the east.

By 3 August the Hverfisfljót had dried up completely. On 7 August, the first visible stream of lava gushed from its gorge, and continued to follow that course for the next two days. The lava slowed considerably thereafter, but steadily piled up in layers until 14 August. By then, however, it had ravaged a pair of sizeable sheep farms that stood on opposite sides of the channel. Now the people of Sída found themselves trapped between two monstrous arms of lava. One, the Skaftá flow, lay to the west and south of Klaustur; the other, the Hverfisfljót flow, lay to the north and east but was stretching its tendrils around to the south. No one knew whether the two arms might close around them, perhaps in the dead of night, in one last fatal embrace.

Dark clouds hung over the district, even as the skies beyond were clear. Ash and sand rained down; lightning flashed, followed by percussive claps of thunder. An all-pervading stench settled across the land. Each arm of lava had its special stink: Jón thought the lava in the eastern canyon reeked of burning wet weeds 'or some such slimy material', while that in the western canyon smelled 'as if burning coal had been doused with urine or another acrid substance.'

Jón, being Jón, was determined to continue his ministries 'no matter how the sparks flew and crackled about me.' One of his immediate concerns was an isolated church farm called Kálfafell, just a few kilometres east of the new lava flow. As one of the few remaining priests in the region, Jón decided that the Kálfafell church and its ornaments – the altar, chalice and other sacred items – were, for the time being, his responsibility.

But he could not rescue the church's treasures until the wind changed direction enough to drive off the noxious clouds. He waited until 14 August and ventured halfway to the church, scouting out whether it was possible to cross in front of the advancing lava flow. By then, however, the lava extended well to the south of Kálfafell and a torrent of water was surging seaward just to its east, making a crossing doubly impossible.

Three days later the lava flow was slowing considerably, although its rushing sound could still be heard issuing from the eastern canyon. Rain showers mixed with ash fell from time to time, but the floodwaters began to subside. On the 20th, Jón decided to risk journeying eastward. As he could not load all of the church's possessions onto his horse by himself, he sought help from among the locals, but most were too afraid to join him. He finally found a willing lad from a nearby farm, and the two made their way mostly without incident, though at one hamlet they became mired in quicksand and were forced to swim the horse from one bank to the other.

Jón then made his usual rounds, looking in on the few souls still living in the shadow of the Hverfisfljót flow. At the Kálfafell church he took as many of its precious objects as he could load onto his horse. To get home he decided to take a different route further inland, thinking the streams would be shallow and easier to ford. The streams were indeed shallower, but so much silt and floodwater had accumulated on the flats that it took the pair fifteen hours to make the crossing. By the time they reached Klaustur they were exhausted and soaked to the bone. No one would ever take that route again, but the church ornaments, at least, were safe.

The volcanic fury kept coming. The first of September saw a second surge of lava emerge from the Hverfisfljót canyon, destroying outbuildings and fields of farms that had stood for centuries. A branch of the flow made its way eastward, engulfing hay meadows and damming a river. On 10 and 11 September, a new surge poured out of the canyon, cascading over the lava that had preceded it. On the 14th, a heavy shower of ash fell through the central part of the highlands. Thereafter, the lava's advance slowed and water once again began flowing through the river gorge. From his home, Jón could see the ruddy glow of distant fires flashing above the mountains at night.

On 26 September, vigorous earthquakes shook the eastern parts of the district, followed by a lava flow so great it dried up most of the remaining rivers and streams. Great columns of smoke and steam rose high into the air. The upheaval continued until 24 October, when powerful earthquakes once again shook the region. The following day, Jón saw 'a great spout of flame' shooting upward, followed by rain mixed with sandy ash. That same day a 'terrible surge of fire' with 'crackling and thudding' emerged once again from the Hverfisfljót gorge.

Although Jón didn't know it, this was to be the last, as well as the most prodigious, surge of lava to come out of the gorge. Lava flowed continuously for five days, filling the lowland area between two mountains. The fire stream then turned westward at a former farm, filling up that valley, and coursed over lava laid down in August, raising it to more than double its previous height.

On Sunday, 2 November, Jón held a service at Kálfafell for those few souls who remained. A north wind brought showers of ash so thick that only the outline of the church could be made out from a distance, but by evening the wind abated to a gentle breeze and the ash clouds dwindled with it. Now it could be seen that the area northwest of Kálfafell, up to the edge of the newly laid lava, was one 'continuous sea of flame'. The light cast by the fires at night was as strong as bright moonlight. The next evening, as Jón and some men set about gathering wood down by the tidal flats, the eerie incandescence illuminated what would otherwise have been a difficult, pitch-black path.

Throughout November lava continued to stream from the Laki fissures, but the flows did not break out of the highlands. Rainfall mixed with ash fell sporadically, and on 24 November a strong earthquake jolted the region. In early December, the fiery glows that had been seen practically every night began to die down, although lava continued oozing from the fissures, which, in turn, produced occasional spouts of steam. A bluish haze, like that seen before the eruptions began in June, crept over the ground, preventing the grass from returning green and strong. The sun and moon, however, were as bright as they ever were, except when clouds of smoke passed over them. Far into the winter, the moon would turn bright yellow in the smoke.

The holidays crept closer, yet no one felt like celebrating. Christmas Eve saw an atmospheric manifestation 'not unlike a work of sculpture', as Jón put it. The weather that day was calm and clear, but a little before sunset a thick cloud drifted over the steep slope behind the Klaustur church and soon took on the form of an oval wreath. As Jón described it, the cloud-wreath had a light blue bulge in the middle, with red, black, yellow, pink and saffron 'branches, curls, and spheres extending out into the wreath itself.' The apparition, seen by many, hung motionless in the sky until it vanished before sunset.

Jón the scientist might have seen this as the product of mineral vapours arising from fresh lava. The priest within him, though, saw it as a harbinger of famine and death. Or, at least, that's what Jón wrote in his account of the Skaftár fires five years later. One wonders how he would have viewed the cloud-wreath if 1784 had brought happier consequences.

The New Year began auspiciously enough with mild weather, but in mid-January a bitingly cold north wind arrived. Snow fell from smoky clouds, accumulating into crusts of ice on the ground; in Jón's church, a pint bottle of communion wine turned to slush. On 7 February the fires were seen in the distance for the last time. This date is usually considered as marking the end of the lava-production phase of the Laki eruption, but earthquakes, clouds of steam, foul odours and occasional ashfall were to plague the region for months. Spring 1784 also saw devastating glacial floods that surged over the sand flats down to the sea. In one such flood, three men were carried away as they attempted to cross a river. Their bodies were never found.

Though the eruptions had eased, the disaster had only begun. Already the people of Sída had lost profitable farms, meadowlands, trout streams and lakes, and salmon fishing along the coast. Now they began losing their lives.

First came the toxins. The volcanic ash destroyed wild plants harvested for food, such as Icelandic moss, angelica root (wild celery), crowberries, blueberries and sea-lyme grass (which was ground up for gruel and bread). The ash also salted the ground with what Jón simply called 'poison'. Nobody knew what this substance was, but within two weeks of grazing in pastures polluted by ash, animals began exhibiting symptoms not seen in recent memory. Horses shed skin, and their tails and manes began to moult. Hard, swollen lumps grew from their joints, especially the fetlocks. Their heads became swollen and their jaws so weak they could barely graze. Unseen, the ingested ash created a caustic acid in the animals' stomachs, which, in turn, corroded their intestines, leading to fatal haemorrhages. Cattle suffered similar afflictions. Large

Sheep farmers try to round up a flock as they walk through a cloud of ash pouring out of the erupting Grímsvötn volcano on May 22, 2011. Ash deposits were sprinkled over the capital Reykjavík, some 400 kilometres to the west.

growths appeared on the jaws and shoulders, and the legs of some animals became so fragile they cracked in two. Hips and other joints were disfigured, often fusing and becoming immovable. Tails and hooves fell off, ribs became warped, and the animals' hair dropped out in patches.

Sheep, the mainstay of rural life, exhibited even more bizarre symptoms. They developed severe bone and teeth deformations in which their jaws became so swollen they protruded from the skin. Incisors fell out, to be replaced by fragile teeth that were studded with yellow and black spots – 'ash-teeth,' as they were called by a farmer who first described these symptoms after the 1693 eruption of Hekla. As the disease evolved, bone spurs appeared in the joints and ribs, along with spinal deformities.

Over time, the wretched animals' molars weakened so that chewing the cud became impossible. When the sheep were slaughtered, the lungs and heart were sometimes found to be distended and covered with pustules. Other times they were shrivelled and inflamed, while some organs had putrefied. Eating such creatures was highly inadvisable, as Jón noted:

> *What passed for meat was both foul-smelling and bitter and full of poison, so that many a person died as the result of eating it. People nevertheless tried to dress it, clean it and salt it as best they know how or could afford to.*

Jón's parishioners soon began suffering from a scourge that he referred to as advanced stages of scurvy or dropsy. The symptoms, however, seem to resemble those manifested in the livestock:

> *Those people who did not have enough older and undiseased supplies of food to last them through these times of pestilence also suffered great pain. Ridges, growths, and bristle appeared on their rib joins, ribs, the backs of their hands, their feet, legs, and joints. Their bodies became bloated, the insides of their mouths and their gums swelled and cracked, causing excruciating pains and toothaches.*

In some cases painful cramps contracted the tendons, particularly at the back of the knee, and there was painful swelling in the hands and feet, as well as the neck and head. Hair fell out. Teeth became loose and often disappeared beneath swollen, bloody gums, whereupon they would rot and fall out. In the most severe cases, the victim suffered putrid sores inside and outside the neck and throat and, in many instances, the tongue 'festered away or fell off'. Other symptoms included shortness of breath, rapid heartbeat and incontinence. In chapter eight we'll see how most of this was probably due to fluorine poisoning.

In the first half of 1784, some 150 locals fell to the ravages of this epidemic, Swedish geologist Carl Wilhelm Paijkull reported in his 1868 travelogue *A Summer in Iceland*. The isolation of some of the farms only magnified the horror. At one farm northeast of Klaustur, all of the occupants were struck down almost simultaneously and lay where they had died until passing travellers discovered their bodies.

By late autumn, the situation for both humans and animals was dire. Laki's ash had spread across nearly all of Iceland, poisoning the landscape far and wide. Huge areas of grassland had been destroyed, and much of the existing hay stocks contaminated. Deprived of pastureland and fodder, sheep and cattle perished. People began starving to death by December 1783, and thousands more of all ages and classes were to follow as winter progressed.

Jón recorded the many appalling measures taken by the starving. Some 'cooked what skins and hide ropes they owned, and restricted themselves to the equivalent of one leather shoepiece per meal, which was sufficient if soaked in soured milk and spread with fat.' Others resorted to cutting up hay into fine pieces and mixing that with meal to make porridge or bread. Fish bones found half-buried along the shoreline were collected, cleaned, boiled and crushed in milk as a gruel. Some in Jón's parish took to eating horsemeat; most of them died. Others, Jón dryly observed, 'would rather die than eat it'.

To relieve the misery of his parishioners, Jón employed his medical background where and when he could. Pigweed and angelica root became a purgative to treat painful diarrhoea and worms. Intestinal ailments were treated with well-boiled curdled milk and whey, porridge, hard-baked rye bread and water, or, failing those remedies, boiled seal meat (without the fat). Swelling of the mouth and gums responded well to warm milk straight from the teat. For swollen joints and contracted tendons, Jón prescribed a dish consisting of dandelion greens thickened with meal, or a 'mercury plaster' or a salve of yarrow and roseroot. Sulphurwort alleviated chest ailments, as did a tea made from thyme. To drive out fetid smells from ulcers, abscesses and other sources, 'it was wholesome to burn bark and juniper wood'.

For all Jón's doctoring, however, people were still dying. That left just one more grim duty for him to oversee: burying the dead. From the time of the first eruptions until the beginning of 1784, the number of deaths was manageable. 'But from the beginning of that year onwards,' he wrote, 'as the winter passed the number grew and grew'. The chapel that had seen the miraculous Fire Mass was turning into a place of death. The winter snows, however, were often so heavy that the dead could not be brought to church. Worse yet, there was not a horse left healthy enough to bear corpses, except one owned by Jón. The animal, stocky and old, somehow made it through winter, though when it wasn't grazing in the poisoned pasture it ate contaminated hay.

In 1784 Jón recorded 76 deaths in his parish. The unprecedented number of corpses created logistical difficulties: it was hard to dig graves in the frozen ground and to find enough space on church land to bury them all. Some weeks, as many as ten bodies collected at the church until the following Sunday, when they would be buried in a single grave. All 76 victims were interred in the same corner of the churchyard, the one spot not inundated by ash and water. For whatever reason, no one had been buried in this section before, and it seemed to the priest as if it had been 'set aside and reserved for this use'.

In keeping with custom, no one was buried on farms or out in the open, except, Jón noted, one man by the name of Vigfús Valdason, who 'cursed practically everything and everyone around him'. Fittingly, it seems, he died alone of exposure on the barren sands of a glacial outwash. His grave was in a nearby lava field, marked by 'a cairn of stones piled over him'.

As the famine extended into spring, people became more demoralized, and social order began to break down. In some

Laki's lava flows today form a rolling landscape covered with soft moss.

houses, there was scarcely anyone healthy enough to tend to the afflicted. In every quarter, it seemed, people were reduced to skeletal vagrants, thieving, begging or dying. Those who were still able to walk fled westward in an attempt to find new land or good fishing, but to do so they had to leave almost all of their possessions behind. Anything not placed in the custody of trustworthy individuals was eaten or stolen by others. Locked stores were broken into and ransacked, and houses and farms looted and burned. Of the 85 farmers who lived in the region before the eruption, only 21 remained. Of the 613 residents, including spouses, children and farmworkers, only 93 stayed behind to eke out a new life.

Into the middle of this calamity, in April 1784, stepped one Magnús Stephensen. The Danish king had dispatched him, a twenty-two-year-old law student at the University of Copenhagen, to investigate the eruption. By birth Magnús was a member of the island's gentry – the son of Ólafur Stephensen, one of the wealthiest and most powerful figures in Iceland, who in 1790 would be appointed governor, the first Icelander in history to represent the Danish government.

After graduating from university, Magnús Stephensen would go on to become a lawyer in Iceland's northwest district and, in 1800, the first chief justice of the Icelandic High Court. But that spring in 1784, he was responsible for being the eyes and ears of the Danish king in a place that had practically been blasted from the face of the earth. His instructions were to collect as much information as he could about the eruption and its lava flows, and to assess its social effects for helping with disaster relief.

According to his official report, Stephensen set out from near Prestbakki on the morning of 16 July 1784, to find the source of the devastation. With his servant and a 'brave old man' he had persuaded to accompany them, he made his way over the moors and toward the blackened edge of the eastern branch of the lava. The little group followed this flow northward until they reached a small mountain, which the lava appeared to have embraced on several sides. Toward the north, he saw smoke issuing from a 'lava stream ... an appearance equally terrible and indescribable'. Squinting through the pall, he saw something else:

We could discern a considerable hillock, or small mountain, greater in its diameter than in its height, whence there also proceeded a thick and black smoke. There I concluded must be situated the source of the eruption.

Stephensen's account, published in Copenhagen in 1785, was one of the first reported eyewitness reports to reach the world outside of Iceland, and thus one of the first to pinpoint Laki itself as the cause of all the suffering. Subsequent researchers, though, have questioned Stephensen's accuracy. Thor Thordarson, the pre-eminent modern expert on the Laki eruption, argues that the report has so many errors regarding the nature and timing of events that it has no practical value. He notes a local rumour that Stephensen never ventured far enough into the highlands to actually reach the source of the eruption.

Stephensen may not have made it all the way to Laki, but he was in southern Iceland when the hardships were at their worst. At the very least, his comments provide a vivid description of the awful reality there:

> *The volcanic eruption having thus been productive of devastation and sickness, both among man and beast, a great famine and unexampled misery throughout the country, naturally ensued. The peasant, who, with the loss of his cattle, was likewise deprived of his sole means of subsistence, and of the best and most valuable part of his property, had nothing else (after having eaten the animals that died by famine and sickness) wherewith to satisfy the painful cravings of hunger, but skins and old hides, which he then boiled and devoured.*

And occasionally, his observations penetrate to the very heart of desperation and the lengths people go to survive:

> *From respect to my readers I forbear to enumerate a variety of other things, which, as articles of food, were in equal or greater degree nauseous and disgusting, and which, were I to detail them, would serve to show what shocking expedients the extreme cravings of appetite will drive men to have recourse to, and how that it is possible to convert almost every thing to food.*

There is no credible evidence that Icelanders resorted to cannibalism, but the hint of unspeakable deeds shows just how hopeless people were in these dark days.

Eventually, Jón himself began to break down. He had lost all his livestock, including his only cow, and his acts of charity on behalf of the displaced had brought him and his family to the brink of destitution. Between 12 August 1783 and late June of 1784, he, his wife and daughters had no milk, butter or cheese, and only contaminated meat to eat and foul water to drink. Now, for the first time, his physical strength was beginning to flag.

Jón decided to travel west and ask for help from the headquarters of the Danish administration at Bessastadir, near Reykjavík. (Today Bessastadir is the official residence of Iceland's president.) There, he was given a sealed moneybox containing 600 rigsdaler,* and instructed to deliver it to the senior official presiding over his administrative district, in the village of Vík. The official would break the seal and oversee the distribution of the money to those farmers most afflicted by the disaster. On his way, Jón stopped overnight at another village where he met the monastic proprietor of Klaustur, who was supposed to help the senior official distribute the money. The next morning, while Jón was out tending to a sick villager, the proprietor broke the seal to the moneybox and removed twenty rigsdaler for himself and eight for a farmer so he could buy a cow.

Now travelling alone toward Vík, Jón met a large group of his parishioners who, impoverished and weakened, were coming west in the hope of acquiring livestock. Without cash, however, their only recourse was to buy them on credit. They had heard that Jón had a great quantity of money at his disposal, and they implored him 'in God's name' to save them from starvation. Given that the seal on the moneybox was already broken, Jón handed them 245 rigsdaler – one or two per man – so they could buy some animals.

* During Jón's time, relatively little money circulated in Iceland. Most business transactions were by barter, especially of fish. When currency was called for, the official Danish rigsdaler was used. A silver rigsdaler was valued at 27 grams of silver and was equivalent to two rigsdaler; a crown rigsdaler was worth three. In Iceland at the time, a sheep normally cost 1 rigsdaler, a cow 7, and a horse 8, though, in the wake of the Laki disaster, these costs rose so that the price for a cow or horse was as high as 10 rigsdaler.

When he arrived in Vík, the district overseer was furious to find that the government seal had been broken, and brought charges against Jón before the Danish governor. Jón pleaded extenuating circumstances. His bishop stood by him, and the case was closed in 1786 with a minimal penalty: Jón was fined five rigsdaler and had to make a public apology for his 'crime'.

The worst blow came on 4 October 1784, when Jón's beloved wife Thórunn, who had been ill for years with kidney disease, died. They had been married for 31 years, and now he found himself alone and with his house in a wretched state. He had no fuel for the lamps and languished in constant, cold darkness. At night, in bed, his hands and feet became swollen from frost. At Christmas he injured his arm and for five weeks could hardly get himself dressed. He began suffering from insomnia and depression, and by the middle of winter was considering suicide.

The entire country was suffering like never before. Deaths had eased off during the summer of 1784, but rose again in the autumn and winter. Failure of the hay crop and the unavailability of pasturage brought renewed starvation to the central and eastern parts of the country during the first half of 1785, while the west and southwest saw another mortality peak in the spring of 1785.

As the summer of 1785 waned, the Danish superintendent revoked orders to distribute money to farmers who desperately needed it. He also decreed that all the homeless people still living in the three districts to the west be forcibly relocated to Sída. The refugees numbered about 40 in all, but it might as well have been 4,000. Food was in such short supply that there was no way anyone already living in Sída could take in one more soul. The only possibility, observed Jón grimly, was to find them a place to die.

These dire days, however, were not to last. During Sunday services on 16 October 1785, Jón led his parishioners in beseeching God 'to send us and these wretches relief'. Afterward, not having much in the way of alternatives, they decided that a small group would travel east to the beaches in search of whatever the sea might provide. One of the men went ahead of the group, accompanied by two boys, to scout out the situation. The rest of the party arrived at the coast on 21 October and, to their amazement, saw that the scouts had killed 70 male seals and 120 pups. It took almost 150 horses to carry this bounty back to Kálfafell, where Jón held a service of jubilant thanksgiving. More than two years after the fires of Laki first arose, this killing of the seals marked an unofficial end to the famine.

The Laki eruption might be looked upon as Jón's moral crucible, in which the fires of the earth tested both his faith and fortitude. He would eventually marry again and rebuild his farm, grateful to be alive. The end, brought on by kidney disease, came on 11 August 1791. Jón was 63 years old. By his own account, he had baptized 309 children, buried 358 of the dead, confirmed 300 souls, married 69 couples, and medically ministered to thousands. This was a respectable enough record for any priest shepherding an isolated parish in an isolated region of an isolated country.

But what Jón couldn't have realized in those final months was how his account of the 1783–84 eruption would change science. Volcanologists still praise his chronicle for its descriptions of explosive activity, gas clouds, and rock and ash falls. It is the only record that describes such a massive lava flow in such detail, including the course the molten rock took down various river canyons. Taken together, such extraordinary detail provides information on the nature and timing of the Laki eruption that otherwise would have passed into history unrecorded.

CHAPTER FIVE

Horrible Phenomena

Europe's 'year of wonders'

ON 10 JUNE 1783, people living in the Faroe Islands, some 450 kilometres southeast of Iceland, witnessed a heavy fall of black ash and acidic rain that scorched the grass and leaves. That same day, an advancing haze appeared in western Norway and northern Scotland. This was the first manifest evidence beyond Iceland that something momentous had occurred. Icelanders were already suffering the consequences of the Laki eruption, but others across the United Kingdom, France and much of the rest of Europe would soon face a different kind of fallout – one they would observe first in naïve wonderment, then in mortal dread.

Like so many factory chimneys, the Laki rift belched its sulphur-rich effluvium high into the air. The emissions were caught up in the eastward polar jet stream flowing toward Europe and then in a series of high-pressure systems. Subsiding air masses carried them back down toward the surface, spreading them in a spiral-like fashion. The result was a caustic fog that curled its poisonous tendrils across the continent.

In Saint-Quentin, northern France, a haze rolled in from the northwest on 10 June; it would persist for about six weeks.

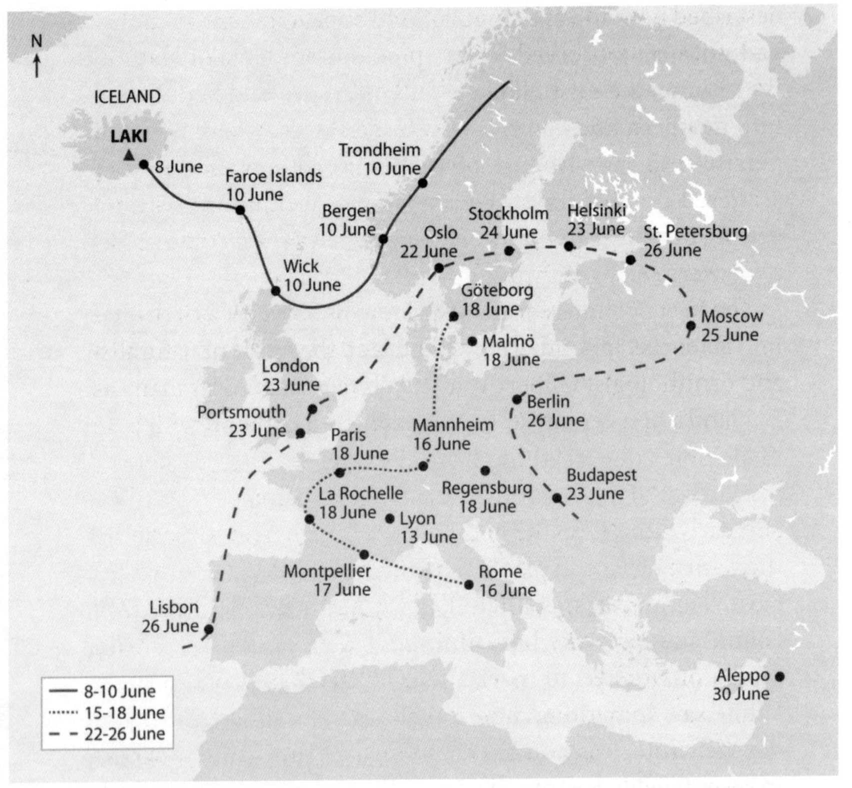

Laki's poisonous haze spread across continental Europe within weeks. Prevailing winds carried it in a spiral pattern across the continent.

Curiously, there was no alarm. Rather, the inhabitants thanked 'Divine Providence that these fogs while stopping some of the sun's rays have prevented the heat from increasing which would have been hard to bear.' Of all the observations of this unusual phenomenon, this is, perhaps, the only one that saw the fog in such a favourable light. Although Europeans didn't know it at the time, Laki was poisoning their entire lower atmosphere.

Reports of the vapour spread wherever it appeared. Some 300 kilometres south in Dijon, a fog, 'by no means a common occurrence', was seen a little before midday on 14 June. In the days that followed, the haze spread into Switzerland, Germany and much of central Europe. In Geneva, reports described how 'fog of a singular kind appeared here, such that had not been observed by any previous students of Nature.' By 18 June, the mist blanketed all of France and had drifted into northern and central Italy. In Le Havre, where it would persist until early August, observers noted how the particles scattered the sun's rays: 'We could look at [the sun] without getting blinded two hours before sunset, as it was then red as if we were seeing it through smoked glass.'

The fog encountered its most eloquent observer as it drifted into Selborne, in southern England, and the home of naturalist and ornithologist Gilbert White. White, regarded by many as England's first ecologist, kept a detailed journal in which he made notes of everything from the weather and the state of his garden, to musings on local birds and animals. In late May and early June 1783, he had made several entries describing the return of several plants that had been damaged by winter frost. 'Honey-suckles still in beauty,' he wrote on 7 June. 'My columbines are very beautiful; tied some of the stems with pieces of worsted to mark them for seed.' But on 23 June, White saw something more perplexing and unwelcome. The day, he wrote, was hot, hazy, and misty: 'The blades of wheat in several fields are turned yellow and look as if scorched with frost.' The following day he dropped in a line from Milton's *Paradise Lost*: 'The sun "shorn of his beams" appears through the haze like the full moon.'

Over the next several days White kept noting the fog's remarkable effect on the sun: 'Sun looks all day like the moon, and shed a rusty red light.' Its unusual appearance alarmed the uneducated locals, he reported: 'The country people began to look with a superstitious awe at the red, louring aspect.' White observed the haze daily between 23 June and 20 July, noting that the wind, no matter from which direction it blew, did not disperse it.

At sea level, where the haze was most noticeable, it took on a bluish or reddish tint and limited visibility to about two kilometres. Many people assumed that it was confined to the lower atmosphere, but, judging by the brightness and colour of the moon and sun, French botanist and meteorologist Robert de Lamanon suspected that it rose quite high. He confirmed this by climbing to the summit of Mont Ventoux in Provence (1,900 metres), where he noted that the haze was still above him. He later learned that shepherds in the Dauphiné Alps (with peaks well above 3,000 metres) had reported that it was perceptible, though more dispersed, at those elevations.

Also telling was how the haze dimmed the light of the stars, a phenomenon astronomers call atmospheric extinction. Barring light pollution, stars can normally be observed in almost every part of the sky, and so they make excellent indicators of a fog's density. Under normal circumstances, the faintest star that can be seen with the unaided eye is rated as magnitude 6. Brighter stars range between magnitudes 5 and 1, with 1 being the brightest. During the nights of densest haze in the summer of 1783, first-magnitude stars could not be seen at low sky altitudes from Scandinavia to northern Italy. In Mannheim, Germany, stars vanished below about 40 degrees (where 90 degrees is directly overhead); in Narbonne, in southern France, stars could be seen rising and setting at about 10 degrees. From locations such as these, the Big Dipper could barely be discerned.

The German-English astronomer William Herschel, who was surveying the sky with his handbuilt telescope from Datchet, near Windsor, noticed a ground-hugging haze on 4 July. He was trying to observe a rich cluster of stars in the constellation Sagittarius which, from England, never rises more than 15 degrees in the sky to the south. In his journal he noted the difficulty in seeing it very clearly: 'The night is extremely hazy near the horizon', he wrote. On 18 August, while estimating the colour of a pair of stars in the southern reaches of the sky, Herschel complained that 'a dry fogg [sic] probably tinges them too deeply'.

There was nothing pleasant about the haze. At times it emitted a sulphurous stench, burned the eyes and throat,

Astronomer William Herschel may have seen the Laki haze as he was trying to observe the night sky from the village of Datchet, near Windsor.

and left a bitter taste in the mouth. It was almost impossible to breathe in such a miasma, and those already afflicted with heart conditions or respiratory difficulties took to their beds. In Franeker, in the Netherlands, physics professor S.P. van Swinden described the situation:

> *On the 24th day it brought with it as a companion a sulphurous odour very readily perceived by the senses, crawling through everything, even closed houses. Men with delicate lungs experienced that same sensation, as if they were turned towards a place in the neighbourhood of burning sulphur. They were unable to contain a cough, as soon as they were exposed to air.*

The environment fared no better. 'The fields showed a very sad appearance', wrote van Swinden; 'the green colour of the trees and plants had disappeared and the earth was covered with drooping leaves. One would easily have believed that it was October or November.' Insects perished atop the blighted leaves.

Suspecting that acid was to blame, some experimenters took things into their own hands. Outside Neuchâtel, Switzerland, paintings left out in a meadow during the fog exhibited strange reactions indeed. Red colours turned to orange, then purple, while black was partially washed out. It matched well what happened when the scientist du Vasquier dunked a painted canvas in diluted hydrochloric acid. 'These observations leave no doubt', van Swinden wrote, 'but that this haze united some acid, or rather some acid gas.'

In some places the haze was accompanied by obvious ashfall. In Caithness, at the northeastern tip of Scotland, 1783 became known as the 'year of the ashie' because of the dark material that rained from the sky. In Italy, a magnet could be used to attract iron particles from the dust that settled over Venice.

One of the haze's strangest attributes was that it could sap moisture from the air. It did not burn off in the sun's rays as a normal fog would. Another unusual trait was how long it lasted: people had seen fogs before, but never one that persisted day after day, for a month or longer.

Added to all these miseries was the intolerable heat that accompanied the haze. Anecdotal descriptions make it plain that 1783's pervasive heat was extraordinary in every way. Gilbert White described it as being so intense 'that butcher's meat could hardly be eaten on the day after it was killed; and the flies swarmed so in the lanes and hedges that they rendered the horses half frantic and riding was irksome.' Writing to the Countess of Ossory from England in mid-July, Horace Walpole described the unpleasantness:

I am sorry your Ladyship has suffered so much by the heat ... Indeed, as much as I love to have summer in summer, I am tired of this weather – it parches the leaves, makes the turf crisp, claps the doors, blows the papers about, and keeps one in a constant mist that gives no dew, but might as well be smoke. The sun sets like a pewter plate red-hot; and then in a moment appears the moon, at a distance, of the same complexion, just as the same orb, in a moving picture, serves for both.

The astronomer Herschel, too, noticed the uncommon heat. On 2 August he recorded hourly temperature measurements in his observing log. The temperature at noon 'in the shade [and in a] free current of air at the back of my gates in the garden' was 90 degrees Fahrenheit (32 degrees Celsius). Not until 3 a.m. did the temperature fall to 59 degrees Fahrenheit (15 degrees Celsius). For southern England, these were above-average readings.

The heat blanketed northern, western and part of central Europe. A letter from Vienna, published in London's *Morning Herald* in September, claimed that 'we have experienced here the greatest heat ever remembered in this country' – as high as 31 degrees Celsius.

The combination of heat and volcanic haze conspired to unleash violent storms. Increased surface evaporation loaded the atmosphere with water vapour, while humidity-loving particles from Laki served as nucleating seeds around which clouds could readily form. Together, these conditions generated violent electrical storms, often accompanied by hail that pounded ships and farms, penetrated roofs and smashed windows. Sudden and violent floods washed soil off the fields. The *Bristol Journal* reported on one such terrifying evening in Leicester in its 19 July issue:

On Thursday night last the inhabitants of this place were alarmed with the most awful appearance of lightning ... exhibiting a wonderful spectacle of dreadful magnificence; before eleven o'clock the whole firmament appeared on fire ... this scene of inconceivable horror continued for near an hour.

The *Sherborne Mercury* reported on 21 July that a storm in London was:

> *violently alarming; the flashes of lightning were remarkably sulphureous, and peals of thunder loud and awful . . . The lightning fell towards the earth, which rendered its effects more alarming. . . As the storm came on in most places the thermometer kept rising. Where the rain fell, the thunder was most violent, where there was no rain the sulphureous stench remained in the air the greater part of the next day, when the heat was more intense than before.*

In some cases, lightning strikes killed livestock and humans. Writing in the August 1783 issue of *Gentleman's Magazine*, a correspondent suggested that there was 'no year upon record when the lightning was so fatal in this island as present.' Storm fatalities occurred all over the continent: the 27 June *London Packet* reported that much of the Corsican town of Saint-Florent 'has been destroyed by fire which was occasioned by a storm of thunder and lightning.' Tremendous storms were reported as far north as western Ireland and as far east as the village of Swidnica in southwestern Poland, where there had been 'so dreadful a storm that there was no distinguishing it from an earthquake.' The freakish weather did not relent for many weeks, lasting, in some parts, until September.

It's no wonder that people became more and more apprehensive as the dreadful summer wore on. In a letter to the barrister Daines Barrington, Gilbert White vividly summed up his thoughts on the remarkable and unpleasant season:

> *The summer of the year 1783 was an amazing and portentous one, and full of horrible phenomena; for besides the alarming meteors and tremendous thunder-storms that affrighted and distressed the different counties of this kingdom, the peculiar haze, or smoky fog, that prevailed for many weeks in this island, and in every part of Europe, and even beyond its*

limits, was a most extraordinary appearance, unlike anything known within the memory of man.

The summer of 1783 was indeed an amazing and portentous one, and the entire year would soon gain the title of annus mirabilis – a 'year of wonders'. In February and March, a series of strong earthquakes had shaken the countryside around Calabria, Italy; at least 30,000 people had died. Some later made a connection between these earthquakes and the summer's mysterious haze, as in the 29 July issue of the *Leeds Intelligencer*:

The foreign papers mention that the haziness, which has lately prevailed here, is general throughout all the southern part of Europe. It is even observed upon the most lofty of the Alps. In Italy, it has occasioned great consternation, as the same appearance of the air was remarked in Calabria and Sicily a little previous to those dreadful earthquakes that have destroyed so many cities. The people of France, too, began to forebode some dire calamity. The Paris Gazette mentions that the churches are most unusually crowded and the shrines of their saints uncommonly frequented.

Linking the haze with the earthquakes was not as far-fetched as it might seem. At the time, one leading theory of earthquakes held that they were born when trapped air rushed out suddenly from the Earth's interior. Indeed, the great Lisbon quake of 1755 had also been preceded by a dry fog, possibly from Katla's eruption that year. On 19 July 1783, the *Norwich Mercury* proposed that 'air received such a concussion by the late earthquakes at Messina and elsewhere, that it became impregnated with sulphurous particles and had all the qualities of lightning without being inflammable.'

On 18 August 1783, a singular event occurred that ramped up the notion of impending cataclysm and added another wonder to the annus mirabilis. That evening, around nine o'clock, countless observers along a 1,500-kilometre path witnessed a brilliant, extraordinarily long-lived fireball arcing across the skies of the North Sea, eastern Great Britain and northern France, emitting an audible buzzing or crackling as it passed. Several minutes after it vanished from view, some witnesses claimed to hear a distant thunderous explosion.

In his report to the Philosophical Transactions of the Royal Society of London, William Cooper, Archdeacon of York, described the scene:

> *The weather, being for this climate, astonishingly hot . . . was sultry, the atmosphere hazy, and not a breath of air stirring ... Toward nine o'clock at night it was so dark, that I could scarcely discern the hedges, road, or even the horses' heads. As we proceeded, I observed to my attendants, that there was something singularly striking in the appearance of the night, not merely from its stillness and darkness, but from the sulphureous vapours which seemed to surround us on every side. In the midst of this gloom, and on an instant, a brilliant tremulous light appeared to the N.W. by N. At the first it seemed stationary; but in a small space of time it burst from its position, and took its course S.E. by E. It passed directly over our heads with a buzzing noise, seemingly at the height of sixty yards ... At last, this wonderful meteor divided into several glowing parts or balls of fire, the chief part still remaining in its full splendour. Soon after this I heard two great explosions, each equal to the report of a canon [sic] carrying a nine-pound ball. During its awful progress, the whole of the atmosphere ... was perfectly illuminated with the most beautifully vivid light I ever remember to have seen. The horses on which we rode shrunk with fear; and some people whom we met upon the road declared their consternation in the most expressive terms.*

Herschel, too, witnessed the passage of the fireball that evening, the light of which 'was so brilliant that I could all the time

Paul Sandby depicted the great meteor of 18 August 1783 as it astonished witnesses on a terrace at Windsor Castle.

distinguish everything around me, almost as visibly as if there had been a moderate flash of lightning of a long continuance.' He noticed the simultaneous haze, which was so thick it interfered with his sky measurements and made bright stars overhead 'barely visible to the naked eye'.

More bright meteor sightings followed over the coming months. Nevil Maskelyne, Britain's Astronomer Royal, noted that after the 18 August apparition, fireballs appeared over England on 26 September and 4, 19 and 29 October. Other tallies later brought the number even higher.

The prevailing explanation for meteors was the one that Aristotle had proposed some two millennia earlier: that they were dry vapours exhaled by the Earth, which ignited near the top of the atmosphere. In 1714, Edmond Halley proposed an extraterrestrial origin, 'a collection of atoms that formed in the ether' that collided with the passing Earth. His theory did not meet with widespread acceptance, and Halley himself later came to reject it, in favour of Aristotelian vapours. Other explanations involved the aurora borealis or some form of electricity. (Today we know that fireballs are the flash of light caused by space rocks burning up in the Earth's atmosphere.)

While the scientific explanations foundered, supernatural interpretations about fireballs were both well established and widely held. The sudden and startling appearances of meteors often induced terror among religious townsfolk, who saw them as signs of God's displeasure or that the world was coming to an end – or both. Newspaper reportage of the summer's extraordinary events likewise tended to suggest that divine forces lay behind the extreme weather and other atmospheric aberrations. In Gloucester, a violent storm on 2 July induced widespread panic. The *Exeter Flying Post* reported on the hysteria that ensued: 'the women, shrieking and crying were running to hide themselves, the common fellows fell down on their knees to prayers, and the whole town was in the utmost fright and consternation.' In a letter to Reverend John Newton on 29 June, poet William Cowper wrote: 'Some declare that [the sun] neither rises nor sets where he did, and assert with great confidence that the day of Judgment is at hand.'

Frightened parishioners near Broué, in northern France, hauled their priest from his bed and forced him to perform a rite of exorcism on the smog. Such social agitation prompted the French astronomer Joseph-Jérôme de Lalande, of the Paris Academy of Sciences, to publish a somewhat condescending report carried by the *Edinburgh Advertiser* to 'quiet the minds of the people.'

> *It is known to you that for some days past people have been incessantly inquiring what is the occasion of the thick dry fog which almost constantly covers the heavens? And, as this question is particularly put to astronomers, I think myself obliged to say a few words on the subject, more especially since a kind of terror begins to spread in society. It is said by*

some, that the disasters in Calabria were preceded by similar weather; and by others, that a dangerous comet reigns at present. In 1773 I experienced how fast conjectures of this kind, which begin amongst the ignorant, even in the most enlightened age, proceed from mouth to mouth, till they reach the best societies, and find their way even to the public prints. The multitude, therefore, may easily be supposed to draw strange conclusions when they see the sun of a blood colour, shed a melancholy light, and cause a most sultry heat.

This, however, is nothing more than a very natural effect from a hot sun, after a long succession of heavy rain. The first impression of heat has necessarily and suddenly rarefied a superabundance of watery particles, with which the earth was deeply impregnated, and given them, as they rose, a dimness and rarefaction not usual to common fogs.

This soothing explanation, however, was challenged by many natural scientists throughout Europe. And there's little evidence that it calmed the minds of the people, especially when they had the chance to read about the year's horrors in so many places.

The story of the amazing summer unfolded at a time when newspapers were proliferating throughout England and the rest of Europe. By 1783 print communication was moving well beyond the old broadsheets, pamphlets and posters advertising theatre programmes, auctions and shipping news. There were now dozens of newspapers in London, and many more circulating throughout England. As summer waned and autumn approached, many of these publications began reporting stories that fuelled a mood of terrible dread. People across England and France, it seemed, were dying in droves from some inexplicable illness, with symptoms that included headaches, eye irritation, breathing difficulties and fever.

In Umpeau, northern France, a priest reported:

Until the beginning of the thaw the parish of Champseru has been afflicted by a pestilential sickness. Patients were afflicted by a sickness of the throat. Many ignorant doctors treated it by bleeding and applying emetics, and after 18 days there were 40 dead. The fogs of May, June, July and August that darkened the sun and turned it red as blood are believed to be the forecast of this curse. May God preserve my parish.

In England, Charles Simeon returned home to Cambridge in the autumn of 1783 to find an appalling scene: 'Many whom I left in my parish well are dead, and many dying; this fever rages wherever I have been.' William Cowper, writing on 7 September, noted that 'such multitudes are indisposed by fevers in this country, that farmers have with difficulty gathered in their harvest, the labourers having been almost every day carried out of the field incapable of work, and many die.' The following day Cowper wrote: 'The epidemic begins to be more mortal as the autumn comes on . . . in Bedfordshire it is reported, how truly however I cannot say, to be nearly as fatal as the plague.' Cowper, who had seen many of his acquaintances taken ill with the 'fever', feared that he was seeing the return of a previous contagion, such as the epidemic fevers that killed some 20,000 in London between 1726 and 1729, or even the Black Death that killed an estimated 100,000 Londoners between 1664 and 1666. Indeed, many people began referring to the mysterious 1783 epidemic as the Black Fever.

We will see in chapter eight just how wrong they were.

As if the heat, storms, meteors and deaths weren't enough to cope with, next came the cold. The coming winter in Europe was to be one of the most severe in the last 250 years, with a mean temperature for January of -0.6 degrees Celsius, more than 3 degrees below the thirty-year average. 'A winter so tedious and severe has never been experienced in this country,' wrote London's *Morning Herald* on 23 March 1784. 'In England men were found frozen to death on roads and in the open country,' *Gentleman's Magazine* noted. 'Great apprehensions were entertained for the poor, who it was feared would freeze to death.' A letter from Edinburgh reported that 'the poor people are in a very distressed condition for want of meal, and many of the sheep and cattle are dying.'

Relief, when it came, was short-lived. In a letter to Reverend Newton in February 1784, Cowper gives thanks for a brief thaw:

> *My dear Friend – I give you joy of a thaw that has put an end to a frost of nine weeks' continuance with very little interruption; the longest that has happened since the year 1739.*

All of Europe was in the grip of extreme weather. In Denmark, 'the winter came so early and so suddenly that the navigational marks in the Sound [between Denmark and Sweden] were lost' as the waterway iced over. In Holland, two skaters travelled 25 kilometres along the frozen North Sea coast. In Sweden, Stockholm recorded its lowest March temperature ever, at -33.7 degrees.

In Vienna, Wolfgang Amadeus Mozart, who had just completed his Piano Concerto no. 14 in E-flat major, found himself distracted by the bitter cold. In a letter to his father dated 10 February 1784, he writes: 'I have just one more question to ask, and this is, whether you are now having in Salzburg such unbearably cold weather as we are having here?' In Vienna, so many people died that the newspaper *Wiener Zeitung* took to publishing lists of the dead from the city and its suburbs. On average that winter, 15 to 25 people died per day, from a total population of just under 210,000, with the largest number – 53 – occurring on 7 January 1784. The newspaper also reported heavy snows that buried people in their homes, and collapsed roofs and city towers.

Major rivers such as the Elbe and the Danube froze over entirely, halting transport across much of the continent. Snow packs built up in the mountains, so that once the spring thaw began, so too did the floods. These turned out to be some of

Ice breakup in Prague flooded the city bridges in spring 1784.

the most devastating in recent European history. In Prague, water levels rose some four metres in just twelve hours. Downstream, in Dresden, a hundred ships under construction were destroyed. In Paris, the Île de la Cité was underwater, and the inspector-general for public health issued brochures on how to safely re-inhabit flooded homes. In the Carpathians, entire villages vanished. The devastation was so great that officials chiselled the floods' record water levels onto bridges and towers along the waterways, so that future generations would know how bad things could get.

The European flooding came with political overtones. Marie Antoinette, the French queen, found herself donating 500 gold coins to flooding victims after she made a thoughtless remark about how snow-bound streets were all the better for her sledging outings. Her husband Louis XVI, perhaps hoping to stave off unrest, ordered government coffers to be opened to aid disaster victims.

Intolerable heat, violent thunderstorms, bitter winters, flash floods: so many disasters seem to come in the year of wonders. Scientists, both professional and amateur, lost no time speculating on possible links. An anonymous account from Klosterneuburg, Austria, drew a connection between the previous summer's dry fog and the fog seen in the subsequent chill:

> *The remarkable fog, which had occurred in the last summer, was also clearly observed in the winter of 1784, except on few days. It was seen in winter during snowfall, as it was always seen in summer during rainfall. From this coincidence we may conclude easily that this special vapour, which had caused extraordinary heat and disastrous thunderstorms in summer, was also responsible for the extraordinary amount of snow and extremely cold temperatures.*

This observer, of course, was pretty much spot on.

The deep freeze was not restricted to Europe. The newborn nation of the United States also experienced an unusually hard and long winter, with the heaviest snowfall ever known in northern New Jersey, prodigious snowstorms in the south, and the longest recorded spell of below-zero readings in southern New England. Ice sealed Baltimore's harbour on 2 January and did not release it until 25 March; at least three ships were lost. During the week of 10 February, temperatures in Hartford, Connecticut, fell below -24 degrees Celsius. There were ice jams in Virginia's James River at Richmond and in the mouth of the Mississippi River, at New Orleans. Ice floes drifted in the Gulf of Mexico.

In Pennsylvania, Henry Muhlenberg's weather diary of 1783–84 is a litany of misery:

December 24: Much snow and bitter cold.

January 1: The New Year set in with an uncommonly deep snow, so that one can scarsely [sic] get out of the house.

January 25: He could not get across the Schuylkill because the streams are too high and full of ice floes.

January 29: Our children would be glad to be back home, but they must wait until there is a path made on the road, for the windstorm has piled up the snow as high as six to eight feet in many narrow places on the road.

February 7: The cold is so continuously severe that it is not to be compared with previous winters.

March 9: Another deep snowfall ... The dreadful winter is setting in afresh.

In Virginia, the Bill of Rights author James Madison tried to make light of the snows in a letter to Thomas Jefferson:

> *We have had a severer season and particularly a greater quantity of snow than is remembered to have distinguished any preceding winter. The effect of it on the price of grain and other provisions is much dreaded. It has been as yet so far favourable to me that I have pursued my intended course in law reading with fewer interruptions than I had presupposed.*

In typical didactic fashion, Jefferson tried to use the extraordinary winter to revive a plan to conduct simultaneous weather observations across the state. (He failed.)

And in Mount Vernon, Virginia, the practical George Washington had his own complaints. Having reached his beloved plantation in time for the holiday season, hoping for a little rest and relaxation, Washington found himself 'arrived at this Cottage on Christmas eve, where I have been locked up ever since in frost and snow.'

Meanwhile, on the other side of the Atlantic, an old compatriot of Washington, Jefferson and Madison was himself shivering outside Paris, and wondering what to make of the sudden and chilling winter.

CHAPTER SIX

The Big Chill

Laki's global fallout

IN 1783, BENJAMIN FRANKLIN was living in Passy, as the United States' main diplomatic representative to France. Although Franklin had other weighty matters on his mind – notably, hammering out a British–American peace treaty following the Revolutionary War – he became fascinated, inevitably, by the oddities of the weather that summer. Franklin was many things – a politician, a writer, and America's first postmaster. At heart, though, he was a scientist.

No record exists of exactly when Franklin first spotted the dry fog that descended over France, but a key document survives that describes how he thought about it. In May 1784, Franklin sent a letter to his friend Thomas Percival, a physician in Manchester, who corresponded with him regularly on meteorological topics. The contents became public that December, when Percival read it in front of the Manchester Literary and Philosophical Society.

Never one to let a natural phenomenon pass him by, Franklin had seized the opportunity to think about the mysterious haze. As usual, he had thoughtful and insightful things to say about where such a fog might have come from – things that cemented his reputation in the history of volcanology. In his letter to Percival, Franklin described the event:

During several of the summer months of the year 1783, when the effect of the sun's rays to heat the earth in these northern regions should have been greatest, there existed a constant fog over all Europe and great part of North America. This fog was of a permanent nature; it was dry, and the rays of the sun seemed to have little effect towards dissipating it, as they easily do a moist fog, arising from water. They were indeed rendered so faint in passing through it, that when collected in the focus of a burning glass, they could scarce kindle brown paper. Of course, their summer effect in heating the earth was exceedingly diminished. Hence the surface was early frozen. Hence the first snows remained on it unmelted, and received continual additions. Hence the air was more chilled, and the winds more severely cold. Hence perhaps the winter of 1783–4 was more severe than any that had happened for many years.

Franklin was far from the only scientist to note the bitter winter of 1783–84, nor was he the first to link it to the dry fog that spread across Europe the previous summer. But in a typical Franklin twist, it appears as if he had been intrigued enough to try a little backyard experiment, to see whether the haze had attenuated sunlight to the point that a magnifying glass (a 'burning glass') could not ignite a fire. Such a minor test would not have been a stretch for a man who famously flew a kite into a thunderstorm to verify the nature of electricity.

In his essay, Franklin goes on to probe where the dry fog might have come from:

The cause of this universal fog is not yet ascertained. Whether it was adventitious to this earth and merely a smoke proceeding from the consumption by fire of some of those great burning balls or globes which we happen to meet with in our rapid course round the sun, and which are sometimes seen to kindle and be destroyed in passing our atmosphere, and whose smoke might be attracted and retained by our

A curiosity for everything: Benjamin Franklin in Paris in 1783.

> *earth: or whether it was the vast quantity of smoke, long continuing to issue during the summer from Hecla [sic] in Iceland, and that other volcano which arose out of the sea near that island, which smoke might be spread by various winds over the northern part of the world, is yet uncertain.*

Here Franklin proposes two ideas, one perhaps more outlandish than the other. The 'great burning balls or globes' he refers to are meteors, such as the fireball that soared across

Europe on 18 August 1783. Richard Payne, a geographer at Manchester Metropolitan University, has argued that Franklin's suggestion is the first time any scientist drew a link between an extraterrestrial impact and climate change.

But Franklin has a second possible explanation for the dry fog: that it issued from a volcano in far-off Iceland. Franklin had never been to that exotic island, but if he thought a volcano might be to blame he would naturally think of the country with the mighty fire-mountains. Of those, Hekla was by far the most famous and among the most active, and so he proposed it as a possible source. But if Hekla was not the culprit, another possibility might be 'that other volcano which arose out of the sea near that island'. That other volcano was Nyey ('New Island'), a short-lived land that passing sailors had seen rising above the waves off Iceland's southwest coast in the spring of 1783. Tales of its dramatic birth would have quickly spread to the continent, and in the absence of any information about the Laki eruption, Franklin was taking his best guesses as to which Icelandic volcano might have produced the menacing dry fog.

Franklin was not alone in his surmises. Although he didn't know it at the time, he wasn't even the first to link the fog to Iceland. That honour probably belongs to French scholar Morgue de Montredon, who presented a paper about the dry fog to a learned gathering in Montpellier on 7 August 1783, more than half a year before Franklin wrote his letter. Morgue de Montredon was a distinguished member of France's national scientific society, and his paper is a detailed and careful summary of the extraordinary meteorological properties of the fog as seen around Montpellier. He concluded that the haze was rich in sulphur and suggested that it originated with a volcano. A volcano probably meant Iceland, and so de Montredon also picked Nyey as the likely source.

Other scholars weighed in independently with similar ideas. Christian Gottlieb Kratzenstein, a physics professor at the University of Copenhagen, blamed the fog on Nyey sometime in the summer of 1783, and so may deserve the acclaim for being the first to link the haze to an Icelandic volcano. But de Montredon was the first to present his ideas in public, and he also had a greater impact among the scholars of continental Europe. Still, more than either of these two, it was Franklin who made the full link between volcanic activity and the ensuing cold winter. Richard Payne notes that unlike the others, Franklin went on to make a wider connection:

> *It seems however worth the inquiry, whether other hard winters, recorded in history, were preceded by similar permanent and widely extended summer fogs, because if found to be so, men might from such fogs conjecture the probability of a succeeding hard winter, and of the damage to be expected by the breaking up of frozen rivers in spring, and take such measures as are possible and practicable, to secure themselves and effects from the mischiefs that attended the last.*

With such prescient words, Franklin's essay brought the idea of a link between volcanoes and climate change into far broader public consciousness.

The aerosols from Laki were lofted so high into the atmosphere that they spread not only across Europe but also, as we'll soon see, across the rest of the northern hemisphere. To understand how Laki could have such a global impact, we'll need to pause for a brief look at some basics of atmospheric chemistry.

The part of the atmosphere that people experience every day – through which we walk, run, drive and fly – is called the troposphere. It makes up most of the mass of the atmosphere: four-fifths of all the nitrogen and oxygen and other elements that comprise the air we breathe is concentrated here. Weather patterns arise, evolve and expire in the troposphere.

Most of humanity's pollution also swarms throughout this lowermost layer.

Now imagine being a balloon rising through the troposphere. As you get higher and higher, the air grows thinner and temperatures drop. Eventually, by the time you reach about -50 degrees Celsius, you get to a point where the temperature stops getting any colder. Now you've reached the top of the troposphere and are starting to move into a drier and more rarefied realm: the stratosphere.

Some of the planet's most important chemistry takes place in the stratosphere. For instance, here lies a layer of the triple-oxygen molecules called ozone, which shields the Earth from the sun's searing ultraviolet rays (when not being depleted by man-made chlorofluorocarbon chemicals). More importantly for our story, the stratosphere is also where long-distance transport can happen. Any material in the atmosphere that makes it past the top of the troposphere and into the stratosphere can travel around the globe much more readily than it can lower down, where it would be washed out by rain and other everyday weather.

So the boundary between the troposphere and the stratosphere is crucial. Get past that point, and you'll be able to stay aloft much longer and travel greater distances. But here's a complicating point: the boundary between those two atmospheric sections, known as the tropopause, varies in elevation depending on latitude. At the equator, the tropopause is typically about eighteen kilometres above the surface, whereas at the poles it is just eight kilometres up. (This difference is thanks to circulation patterns in the atmosphere, particularly jets of air known as the subtropical and polar front jets.) Therefore, gases and aerosols ejected by volcanoes at tropical latitudes, such as in Indonesia, have a lot further to go to reach the stratosphere. In contrast, those from volcanoes closer to the poles, such as in Iceland, have a shorter distance to travel and can more readily penetrate the stratosphere.

Volcanoes eject many things – including ash, rock and gases – but the main factor that affects climate is the amount of sulphur. It initially billows out of volcanoes as sulphur dioxide gas, but within about a month this oxidizes in the stratosphere, or combines with other compounds, to form sulphuric acid. These acid vapours, along with water vapour, then condense into sulphate aerosol particles, which are the major player in volcanic climate change.

Lower down, in the troposphere, rain washes out most sulphate aerosols within a matter of days. But particles that make it up into the drier stratosphere may survive for several years, plunging back down to the surface if they get mixed into descending air masses at mid-latitudes. They can also get sucked down through atmospheric vortices at the planet's poles. Both processes take time, which means the volcanic particles can influence climate long after the ash plume has faded.

Most eruptions never fire their aerosols high enough to reach the stratosphere, but those that do can affect climate in several ways. For one thing, the particles may warm the stratosphere by absorbing sunlight. They can also accelerate the rate of chemical changes, like the ozone depletion that happens when chlorine particles break apart ozone molecules. (Pinatubo's 1991 eruption, in the Philippines, temporarily reduced ozone by as much as 20 per cent at certain layers in the atmosphere.) Most importantly, stratospheric particles can act as a giant sunscreen: they are just the right size to scatter incoming sunlight back into space, cooling the ground underneath as a result. In a general sense this observation is nothing new: after the 44 B.C.E. eruption of Etna, Plutarch observed how the haze of particles spewed from the eruption temporarily dimmed the sun. Modern science, though, has greatly improved our understanding of how this happens.

A volcano's climate-cooling power thus depends on two main factors: how high its gases are injected into the atmosphere, and how many sulphur aerosols are produced. Eruptions of around the same magnitude on the VEI scale can have very different climatic effects. The 1980 eruption of Mount St. Helens, for instance, had relatively little sulphur dioxide in its plume and cooled the planet very little, while the similarly sized El Chichón eruption in Mexico, two years later, cooled the planet quite a bit since it was so sulphur-rich. And that's nothing compared to the sulphur giants. Pinatubo put about 20 million tonnes of sulphur dioxide into the stratosphere. But Laki spewed out more sulphur dioxide – about 122 million tonnes of it – than any other eruption in the past 1,000 years. That's more than enough to wreak climate havoc well beyond Iceland and the rest of Europe.

Just how much of the planet Laki affected is perhaps the biggest unanswered question about the eruption. There's no doubt that the haze itself travelled far afield. Winds blowing toward the east spread it to Africa, the Middle East and beyond. By 1 July the dry fog masked the sky above central Asia's Altai Mountains, some 7,000 kilometres from Iceland. It may have even spread to central China, as chronicles for that year from Henan province describe a 'severe dry fog – sky is dark'.

Whether the volcanic haze actually spread all the way to North America is controversial. Benjamin Franklin asserted that the fog had been seen over much of the continent, and a missionary in eastern Labrador reported that the air was:

> *filled with the finest smoke so that the sun shone completely pale . . . It is now known to be sure that this smoky air which has occurred in the summer of 1783 over nearly all Europe, has found its origin at the earth fires in Iceland at which possibly the earthquakes at Calabria might have contributed . . . It seems this fog has occurred over the whole northern hemisphere, if not further.*

Few records exist, though, of any haze in northern American cities.

Haze or not, Laki almost certainly did have an effect on North America. Tree rings in northwest Alaska show that trees packed on less wood in the summer of 1783 than in any other year for four centuries. Stories passed down among the Kauwerak people also suggest that a disaster (the Inuit know it as the 'time that summer did not come') may have decimated the population around that time. One tale describes a woman who left her village full of the dead and journeyed with her baby for hundreds of kilometres along the coast, scavenging bits of roots and dried fish along the way until she found other survivors. The terrible summer is described as proceeding pretty much as usual until late June, when suddenly temperatures plunged, snow returned, and the ground froze all the way through to the next year's spring. There's no way to be sure this all happened in 1783, but it makes sense that the hard times would have occurred in the same year that local tree rings recorded dire environmental conditions.

More controversial is the idea that Laki may have stretched its tendrils all the way to South America. In 2010, a Portuguese scholar reported that the astronomer Bento Sanches Dorta had recorded higher-than-average incidences of dry fog and haze in Rio de Janeiro in the autumn of 1784. 'The months of September, October, and November were dominated by a certain kind of fog, or dense vapour, that obscured the Sun during the day and the stars by night,' Sanches Dorta wrote. This haze might be traceable to Laki, which would make it the first record of any Laki impact in the southern hemisphere, but there's little in the way of supporting evidence – and Sanches Dorta himself speculated that the fog came from a submarine volcanic eruption not far from Brazil.

For modern scientists, the challenge is to understand exactly how Laki's dry fog caused such dramatic climate change across the northern hemisphere. Overall, the Laki eruption emitted about as much sulphur dioxide as 12,000 coal-fired power plants do in a year. But not all of this climate-altering chemical spewed out at once: it came in pulses spread out over eight months, with most of it arriving during the first six weeks of the eruption.

Laki was particularly sulphur-rich because of the magma that fed the eruption. In 1996, Thor Thordarson and his colleagues published a seminal paper that used geologic detective work to hunt down the chemistry of the original magma. By looking at tiny bubbles within rocks erupted from the Laki fissure, the scientists calculated the chemical makeup of the original magma. They discovered that Laki was particularly efficient at separating out sulphur and other volatile elements (such as chlorine and fluorine) from the molten rock, and spitting them into the air.

Gases separate out of magma as it rises toward the surface, depressurizing along the way. Laki's magma rocketed upward with the force of a jet engine. By the time anyone in Iceland realized what was happening, Laki was fountaining fire more than 1,400 metres into the air. This suggests the magma was gushing out at up to 170 metres per second – maybe even twice as fast in the early stages of the eruption. At these speeds, Thordarson's team estimated, half to three-quarters of the gases contained in the magma would have separated out.

But how high did Laki's sulphate aerosols reach? Were they really high enough to penetrate the stratosphere and thus be transported around half the planet? Most experts on the eruption, including Thordarson, think the Laki plume was high enough to reach the stratosphere. Analysis of eyewitness reports suggest the plume must have risen at least nine kilometres and probably went as high as thirteen kilometres. At Laki's latitude of 64 degrees north, that would place it just past the tropopause and into the lower stratosphere.

This claim gets some solid support from ice cores drilled into the Greenland ice sheet. Every year, snow falls atop Greenland in winter and melts a little in summer, compacting and building up layer after layer of ice. Gas bubbles trapped within the ice preserve ancient air, which researchers can use to trace changes in the planet's atmosphere over time. Chemical elements within the ice itself also provide clues to other environmental changes, from the rise of industrial pollution to wildfires.

Several international expeditions have drilled cores from Greenland's ice sheet, and many of those cores contain layers rich in volcanic sulphur. Most such layers can be traced back to particular eruptions, such as Krakatau in 1883. Others are mysterious peaks of sulphur with no known volcano to blame them on. Ice layers dated to the year 1258, for instance, show a huge spike in sulphur that implies a volcano eight times as big as Krakatau must have gone off around that time — though no one knows which volcano that might have been. The ice core record of Laki, however, is clear.

Many of these precious ice cores are preserved in Denver, Colorado. Inside a huge building at the Denver Federal Center – once the largest warehouse west of the Mississippi River – sits the National Ice Core Laboratory. All federally financed U.S. drilling projects send their ice cores to be stored here. It's sort of a frozen library, with row after row of silvery one-metre-long core tubes taking the place of books on floor-to-ceiling shelves. In this case the library represents a precious storehouse of information about past climate.

On a cloudy February morning, we make our way to the ice core lab to see the remains of Laki for ourselves. Our guide is chief curator Geoff Hargreaves, who trained as an oceanographic technician and spent years escorting complex scientific equipment on and off research vessels. Now he puts his talents to work creating new ways to prepare and store ice cores for posterity.

Before seeing Hargreaves' ice, though, we have to bundle up. On go a pair of thick coveralls, protective 'bunny boots', a warm hat and gloves. To enter the freezer itself, we go through a series of heavy doors, with increasingly dire warnings about the temperatures ahead. The main core storage is at a bone-chilling -38 degrees Celsius – the kind of temperature that renders people what Hargreaves calls 'cold-stupid'. Reflexes slow down, talking seems difficult and thinking just as arduous. We're glad to have someone experienced in deep cold to escort us through.

Quickly we stumble into the 'warm' room that clocks in at only 24 degrees below. With high white walls and various preparatory tables sitting around, this is the area where technicians and scientists actively work with cores. It's cold enough to keep the ice safely frozen, but those extra few degrees above the deep-freeze temperature is enough to make working here simply uncomfortable rather than dangerous. Some workers can stay in the 'warm' room for up to four hours; Hargreaves says he usually bails after about thirty minutes. For our purposes, the sooner we see the Laki core the better.

Hargreaves leads us over to a black-curtained alcove, which looks something like a photo booth at a train station. He pulls aside the drape and the first thing we see is Arnold Schwarzenegger staring back at us. A lifesize cut-out of the actor, dressed as Mr. Freeze from the movie *Batman & Robin*, is on guard here. Mr. Freeze watches over whatever ice core happens to be placed on the metal tray in front of him – which today is a half-core of ice drilled by the American GISP2 project in the early 1990s. The Denver lab has 3,000 metres of ice cores from this site in Greenland, starting at the top of the ice sheet and going down to bedrock. We are interested in only the uppermost part of the core, the part that holds ice just over 200 years old. The particular segment Hargreaves lays out for us once rested between 71 and 72 metres deep. It lies in front of us, a half-moon of ice a metre long, formed as the American Revolutionary War was ending and the first balloonists were taking to the skies over France.

But it's really hard to see the remains of Laki here: it all looks like pale blue ice to us. Fortunately, Hargreaves has already passed this core through a machine that measures electrical conductivity, and he can see that something very odd happens about 0.095 metres into it. So we look at the

The deep freezer of the National Ice Core Laboratory, near Denver, Colorado, contains polar environmental history stretching back hundreds of thousands of years.

core's left edge and scan our eyes to the right by about ten centimetres. And there it is, barely visible but corresponding to the spike in conductivity that Hargreaves has measured: a dusky band a few centimetres wide crosses the core. It looks as if someone had spilled a bit of cigarette ash on the ice and forgotten to clean it up.

We are looking at some of the particles that Jón Steingrímsson would have seen rising as black clouds above the hills behind Klaustur. The particles danced upward and were caught in the great circulation vortex over the Arctic. Eastward they went, sweeping across Scandinavia and northern Asia and then circling all the way around again, to settle out in snow falling over central Greenland. Year after year the thin ash layer got buried, compressed, and crushed down in the ice core until scientists, finally, retrieved it.

We know that the Laki eruption belched a lot of material high into the air, and that this material travelled a very long way. The question that now faces us is this: how did those emissions cause the freakish weather of 1783–84, and to what extent did other meteorological patterns also play a role? For answers, we need to turn to some of modern science's most sophisticated tools: computer climate models.

Computer models of volcanic eruptions can be used to predict where an ash cloud might spread, making it possible to plan the evacuation of residents, or the clearance of air-space. Models developed for volcanoes can also shed light on what could happen if other unwanted particles were to spread throughout the atmosphere, such as those created by a nuclear explosion or a large-scale blast of pollution.

Few scientists have thought as much about the potential of climate modelling and volcanoes as Alan Robock, of Rutgers University in New Jersey. With twinkly eyes framed by a fringe of white hair and beard, Robock looks like a jovial Santa Claus – until he starts talking about the physics of nuclear destruction. A former Peace Corps volunteer who visited the Soviet Union during the Cold War, and Fidel Castro during the US trade embargo with Cuba, he's not shy about questioning the social relevance of his research.

For Robock, volcanoes are natural laboratories for exploring the consequences of disturbing the planet, and climate models are his main tool. The year after the 1980 eruption of Mount St. Helens, he published a paper in *Science* explaining why it would have little to no global climatic effect; he turned out to be right. He has also helped elucidate many of the mechanisms by which Laki changed temperatures over half the planet.

In a 2006 paper written with Luke Oman and others, Robock combined a popular NASA climate model with another that specialized in atmospheric sulphur chemistry. The scientists first tested the approach by loading it with information about sulphur released by the 1912 Katmai eruption in Alaska and the 1991 Pinatubo eruption in the Philippines. When the model was run, it correctly showed where the aerosol clouds from those volcanoes had spread. The researchers then re-started the model with information about the Laki eruption: what time of the year it occurred, and how much sulphur erupted from the fissures. The model assumed that the erupted particles would have been injected some nine to thirteen kilometres high.

Calculations showed that Laki's sulphur dioxide emissions would have reached their peak in late June 1783 and then converted to sulphate aerosols over the next few weeks. By late August the atmosphere's sulphate load would have maxed out, after which the material would have drifted all the way around the northern hemisphere.

Up in the stratosphere, Laki's particles began their climate-changing work. By absorbing outgoing radiation from the ground, they began warming the air around them. By reflecting incoming solar radiation, they began cooling the planet's surface below. (Despite the extraordinary heat in Europe, for most of the rest of the hemisphere that summer was a chilly one.) Those changes, in turn, would have triggered a cascade of other modifications to the atmosphere, as weather patterns shifted around the globe in response to this new climate forcer in their midst.

In a follow-up paper to the original Laki modelling, Oman, Robock and colleagues took a step further to find out what had happened next. Once again they fed information about Laki into a computer model, then looked to see what it told them about changing atmospheric conditions and how those affected the African and Indian monsoons, crucial weather systems that provide desperately needed rainfall to millions of people.

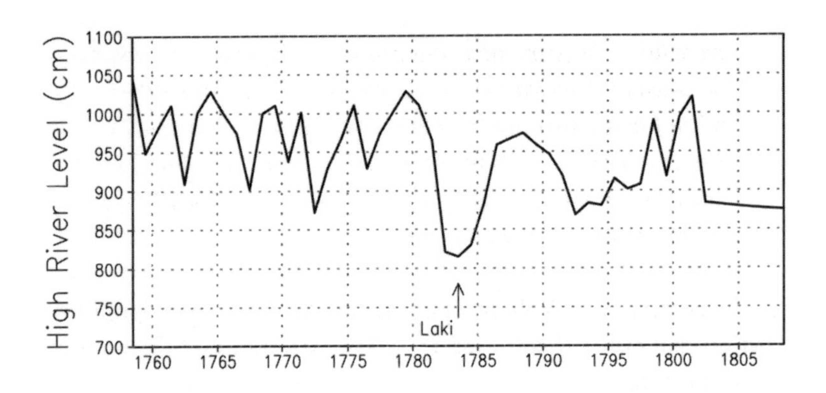

Laki's haze played havoc with temperature and weather patterns, causing levels of rivers such as the Nile to drop precipitously.

Oman's team showed precisely how Laki would have set off a devastating chain of events. Normally, differences in temperature between the land and the oceans set up strong wind patterns that allow monsoons to develop seasonally. But Laki's eruption cooled land masses in the northern hemisphere significantly, by one to three degrees Celsius. Suddenly the land was not all that much warmer than the ocean, and the monsoon didn't have much surface heat to fuel its winds. In Africa in particular, the monsoon simply failed to materialize in the summer of 1783.

With no monsoon, Africa began to dry out. In the western part of the continent, the level of the Niger River began to drop. More importantly, to the east the Nile, too, began to dwindle. For millennia, farmers eking out their livelihood along the Nile had relied on the mighty river's annual flood to replenish and irrigate their lands. That summer, the life-giving floods never came, and neither did they arrive the following summer. With no water, crops failed and famine ensued.

Travelling through northern Africa, French nobleman Constantin de Volney wrote of the disaster:

Soon after the end of November, the famine carried off, at Cairo, nearly as many as the plague; the streets, which before were full of beggars, now afforded not a single one: all had perished or deserted the city ... Nor shall I ever forget that, when I was returning from Syria to France, in March 1785, I saw, under the walls of ancient Alexandria, two wretches sitting on the dead carcase [sic] of a camel, and disputing its putrid fragments with the dogs.

By January 1785, Volney reported, one-sixth of Egypt's population had either died or left the country because of the failure of the Nile.

Beyond Africa, Laki's climatic effects are trickier to trace. One reason for this is that many confounding climate factors were at play in 1783–84, such as El Niño. The El Niño Southern Oscillation is an occasional climate pattern in which the eastern Pacific warms up while the central and western Pacific cool down. (Its name means 'the boy child', a reference to the birth of Jesus, as El Niño often makes its first appearance around Christmas off the coast of Peru.) The pattern repeats every two to seven years, affecting weather around the globe. An El Niño was well underway in 1783, a fact that complicates efforts to look for wider climatic effects from Laki.

India, for instance, suffered droughts and famine in 1783 that may have killed up to eleven million people. But this climate aberration might have been driven at least partially by El Niño and not by the Laki eruption. The same may be true in Japan, where the late summer of 1783 saw unusually cold temperatures along with heavy rains. The combination drowned rice paddies, leading to one of the worst famines in Japanese history, in which tens of thousands of people may have perished. Japan's story is also complicated by the fact that its own volcano, Mount Asama, erupted for three months starting from 9 May 1783. Tens of thousands of people died in ashflows and mudflows from this eruption. (Unlike Laki, Asama doesn't seem to have injected enough sulphur into the stratosphere to make a global impact.)

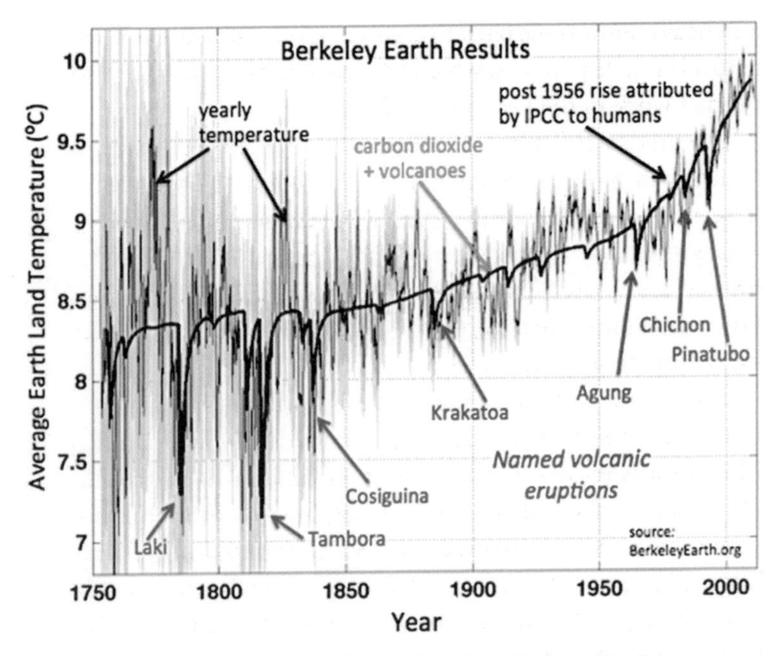

The cooling effect of volcanic eruptions can be seen in this chart of land temperatures from 1750 to the present.

In China, too, parts of the country had an unusually cool summer in 1783, and limited data suggest that deaths spiked soon thereafter. Again, however, it remains difficult to tease out the effects of a severe El Niño year from sulphur-loading in the atmosphere due to Laki.

In fact, one Laki expert thinks that the volcano should be entirely absolved of responsibility for the cold winter of 1783–84. Rosanne D'Arrigo, a tree ring expert at the Lamont-Doherty Earth Observatory outside New York City, points to the winter of 2009–10 for an analogy. That winter was one of the coldest and snowiest ever recorded in parts of western Europe and eastern North America. February blizzards in Washington DC had locals proclaiming that 'Snowmageddon' had arrived, while in the United Kingdom newspapers reported on the 'Big Freeze' that blanketed the country in white from the Isle of Skye to the English Channel. The frigid temperatures and heavy snow trace back to an unusual combination of two natural climate patterns, which D'Arrigo thinks may have also been at work in the year after Laki went off.

The first such pattern, the North Atlantic Oscillation, is a variation in surface pressure that regularly causes temperatures to seesaw across the North Atlantic. When the oscillation is in what's known as its negative phase, temperatures in western Europe and eastern North America are usually colder than normal. At the same time, Canada and Greenland see warmer than usual temperatures.

The second pattern is El Niño, which typically brings more rainfall than usual to certain regions. In 2009, a fairly strong El Niño was locked in place. Essentially, the North Atlantic Oscillation provided the cold to London and Washington, while the El Niño provided the wet. D'Arrigo and her colleagues have used tree rings to reconstruct the North Atlantic Oscillations and El Niños over the past 600 years, and in 2011 they reported that a strong combination of the two caused the chilly temperatures during the winter of 1783–84. Which may mean that Laki is not to blame.

Once again, the key question is how high Laki's aerosols travelled and how long they persisted in the atmosphere. To D'Arrigo, most of the aerosols would have washed out a few months after the first violent eruptions. Others disagree. Atmospheric modeller Anja Schmidt, at the University of Leeds, has calculated that the Laki aerosol cloud would have circulated long enough into the autumn to definitely contribute to winter cooling. And various other records, including additional tree rings, show that cooling lasted across the northern hemisphere for up to three years after the eruption. That's far too long to be explained by a combination of the North Atlantic Oscillation and an El Niño.

The final complication in understanding Laki's climatic effects is the fact that the eruption took place during an extended cold spell known as the Little Ice Age. There is no absolute agreement as to when this era began and ended, but it's widely accepted that Europe started to cool down in the early part of the fourteenth century, and that temperatures began to rise again in the middle of the nineteenth century. The continent endured frequent spells of severely cold weather during this period, and climate-related disasters occurred several times: the 'Great Frost' of 1740, for example, devastated harvests and led to a famine in Ireland just as bad as the more famous one a century later.

The Little Ice Age was a complicated phenomenon, and it's not entirely clear what caused it. Many scientists attribute it in part to an extended period of low solar activity, as if the Sun's thermostat got stuck on low and stayed that way for hundreds of years. With less sunlight arriving at Earth, temperatures would have dropped. But that can't be the whole story, because European temperatures fluctuated wildly throughout the whole of the Little Ice Age, driven by a complex interplay between all facets of the Earth's weather systems. And it's possible that volcanoes played a role in kicking off the Little Ice Age. A 2012 study, led by Gifford Miller at the University of Colorado, proposed that a five-decade-long spurt of eruptions, beginning in the mid- to late thirteenth century, could have triggered a planetary chain reaction that affected sea ice and ocean currents in a way that abruptly lowered temperatures, and kept them low.

As Robock and others have shown, the eruption of Laki almost certainly disturbed the atmosphere enough to cause some amount of climate havoc in the years following 1783. Yet climate modellers can't explain why the summer of 1783 was so hot. Those who watched the haze descend across Europe often noted that the warmest days seemed to be correlated with the days of the thickest dry fog. But scientists can't reproduce this effect in their climate models. To Robock and others, this puzzle remains the greatest unsolved mystery of Laki. A sophisticated model that can parse the exact amount of rainfall over the Nile for months on end cannot explain why summer temperatures would have been broiling over Europe.

Such open-ended issues complicate the calculation of Laki's final death toll. The official count, according to the authoritative reference work *Volcanoes of the World*, is 9,350. But tens of thousands more may have perished across Europe from the effects of breathing in the particles day after day, and if you add in the famines in Egypt and possibly Japan, Laki suddenly becomes a much bigger killer. Pressed for a complete tally, Thor Thordarson suggests that more than 1.5 million people may have lost their lives as a result of the eruption. John Grattan, a geographer at Aberystwyth University who has done the most work on Laki mortality rates, speculates the death toll may even have been as high as six million.

In perhaps the biggest stretch, some environmental historians have even argued that the Laki emissions may have spelled the end for French royalty. The harsh winters, cool summers and heavy rains set up a series of crop failures across France throughout the early 1780s. Bread became scarce, and peasants became angry and desperate, particularly after a drought in 1788. The following year, revolution broke out. Few people would accept this as a straightforward case of environmental determinism: complex socioeconomic and political factors were at work in the French Revolution. That said, Laki clearly disrupted life in France for years on end. Those storming the Bastille might not have known anything about the Icelandic volcano, but perhaps it was an unseen player in the events of 1789.

CHAPTER SEVEN

Laki Today

Life in the mountain's shadow

If you want to visit the craters of Laki, the first thing you need is a proper vehicle. F206, the overland path that leads from the heavily travelled ring road north toward Laki, is one of Iceland's infamous 'F roads' – unpaved dirt tracks that sometimes vanish altogether over lava highlands or under roaring rivers. For this kind of terrain, a rental car just won't do. You need someone like Trausti Ísleifsson and his jacked-up four-wheel-drive van. Trausti and his brother Gudmann run an adventure company in Klaustur, and they unhesitatingly agree to take us to the Laki craters even though we are visiting at a time – mid-June – when the F206 track is often still buried by the winter snows. Fortunately, the spring of 2012 is warm enough to clear a path to the craters. So right after breakfast on a Wednesday morning, Trausti rolls his white van on its massive knobbly tyres up to the front door of our hotel. He is the quintessential Icelandic tour guide: tall, blond, with flawless English, and kitted out in rugged and expensive-looking outdoor gear. We are his only passengers.

Laki is just thirty-five kilometres from Klaustur as the crow flies, but getting there and back is a lengthy affair. On a drizzly,

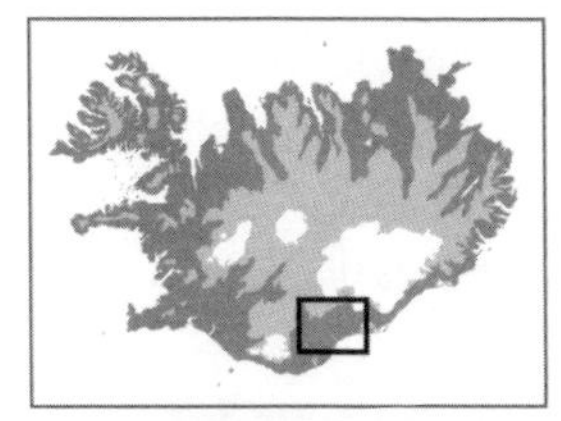

The farms and hamlets along Iceland's south-central coast lie directly in the line of fire downslope from Laki's craters.

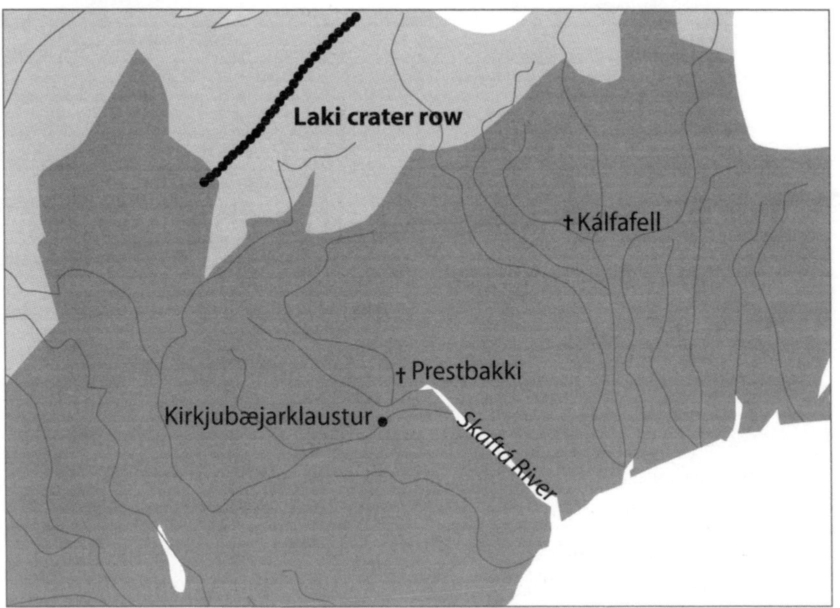

cloudy morning we start by driving about six kilometres west from town, through an eerie hummocky wasteland. What look like soft pillowy forms are actually hard black rocks carpeted by pale green and gray arctic moss. They are the cooled remains of lava from the eruption of 1783–84. The strange lumps stretch on both sides of the ring road, for nearly as far as the eye can see. Forever useless to farmers, this land is now home only to birds.

As we turn off the ring road onto F206, Trausti pulls over the van to let air out of the monster tyres. As we start rolling again, we immediately see the wisdom of this move. The road is unpaved and boulder-strewn, beset with axle-breaking potholes, and the lower tyre pressure allows us to navigate the obstacles better. This early in the season, hardly anyone is on the road except us and one unfortunate-looking Subaru sedan that creeps along, scraping its underside on the rough track. We pass it, wondering what it will do when it encounters the swiftly flowing rivers. Trausti, of course, barrels through fearlessly with just enough clearance under his van. He even halts halfway into the river to dip his water bottle into the icy, clear meltwater. 'You won't taste anything better,' he tells us.

The landscape is classic Iceland: rolling fields of black lava misted with the green of low-growing tundra plants. Pockets of lingering snow nestle into the sides of hills and ridges, and from time to time we spot the massive bank of ice that is the Vatnajökull ice cap glowering in the distance. Out here, the only sign of civilisation is a single blue sign marking the entrance to Vatnajökull National Park, one of the country's three national parks and the biggest in all of Europe.

A ceiling of gloomy clouds extends from horizon to horizon, but even those can't dampen our mood as, an hour and a half after turning onto the track, we finally approach Mount Laki itself. As if on cue, two whooper swans, startled by the van's sudden appearance, take wing. Nothing except them and the clouds seems to be moving out here.

Then Trausti points ahead. After all the anticipation, our first glimpse of Mount Laki is a little less than impressive. It's just another black ridge of rock among other black rocks. Like many of Iceland's mountains, it formed in a subglacial eruption in the distant past when magma erupted under ice, cooling quickly and turning to rock. Today Mount Laki is a craggy, weathered mound with an elevation of 818 metres, comparable to other peaks in the region. But we're here because the mountain sits smack in the middle of the volcanic fissures we came to explore. If you want to see the crater row, this is where you begin.

Trausti brings the van to a halt in a small black sand clearing beneath yet another lava ridge. He waves us out and lights up a cigarette. A bored-looking park ranger cautions us not to stray off the path or pick tundra flowers. Then we are off to climb Mount Laki by ourselves.

The hike is straightforward but steep, so we keep our heads down as we clamber and scramble along the lava trail. An occasional glance over our shoulders at the receding parking area, more scrabbling, a brief rest, a final push, and then we are at the top.

The view is stunning in all directions – lakes, mountains, glaciers – but what we've come to see lies on either side of the windswept summit. A single file of volcanic craters stretches all the way to the horizon, their moss-covered flanks and swales dappled, here and there, with patches of snow.

From a vertiginous shelf, we look off to the southwest: these are the older craters, the ones that opened up on 8 June 1783, and in the following weeks spewed the thousand-metre-high fire columns that Jón Steingrímsson and others saw above the hills behind Klaustur. Lava from these craters surged down the Skaftá gorge and spread out upon the lowlands, destroying farms and threatening Klaustur itself on the Sunday of Jón's Fire Mass.

We turn the opposite way and gaze towards the northeast. Yet more craters stretch off into the distance and disappear below a low cloud bank that hovers above icy Vatnajökull, brilliant white in full sunlight. These craters were the later ones to erupt, beginning the last week of July 1783. They sent lava flowing down the eastern path towards Klaustur and its weary and dying villagers.

This perspective is the only way to really visualise the fearsome power that once ripped this landscape wide open. Laki is but a name, an abstraction, until we see this mighty gash in the earth. Its beauty belies the devastation it once unleashed. How could a land so tranquil and green have once been a terrifying inferno? We take our photos, pile a few stones on a summit cairn, gaze on the crater row one last time, and head back down.

Seen from atop Mount Laki, the crater row stretches off to the Vatnajökull ice cap in the northeast. In 1783, this scene would have been a row of flames.

Later that afternoon Trausti takes us to walk through several of the Laki craters. We drive slowly south and west, marvelling at the colour and variety of forms lava can take, from blue-black pillow-like lobes to reddish sharp-edged rocks. Many are shot through with vesicles, the holes left by gas bubbles as the lava cooled and hardened to rock. Some flow features in the lava resemble brown toffee that has yet to harden.

We come across deep green mirror lakes, gentle burbling watercourses and, here and there, clumps of ground-hugging wildflowers – golden root, purple moss campion, white rock-cress and yellow meadow buttercups. Everywhere sprawls a carpet of that grey-green Icelandic moss, giving the otherwise coarse landscape a soft, impressionistic appearance. As sun-light alternates with a gentle mist that occasionally turns to snow, we almost forget we're walking through what was once a death trap.

ꝏꝏꝏ

After the long drizzly day, we're happy to return to the warmth of the Hotel Klaustur, a modern building just a few steps from the centre of town. Or, rather, a few steps from the traffic roundabout and petrol station that serve as the centre of town. Home to fewer than two hundred people, Klaustur is more a way station than a destination for tourists, and survives mainly because it is a convenient stopping place between Vík, the coastal town to the west, and the grand scenery of glaciers to the east.

Just as it was in the eighteenth century, Klaustur remains a farming community, though farms are being abandoned as young people move to Reykjavík for jobs and other opportunities. The main part of town lies against ancient sea cliffs that rise forty metres to a large lake above. Sheep seemingly defy gravity as they make their way along the cliffs on narrow grazing paths.

The broad Skaftá River flows west to east along the southern edge of town, a clear burbling ribbon that calls out for a fly fishing rod. Before the eruption you would have had to cross it by boat but today you can practically wade across, thanks to the lava that filled up the gorge this river once flowed through.

Other main attractions are a strikingly forked waterfall called Systrafoss to the west of town, and a natural pavement of basalt columns known as the Kirkjugólf or 'church floor'. But we're here for more than a quick visitor stop. We've come to interview a local community leader, the caretaker of Jón Steingrímsson's legacy.

We meet Jón Helgason in our hotel's elegant restaurant. It's hard to believe he's in his eighties: his greyish-white hair may be thinning away from his broad forehead, but there's still plenty of it. His eyes are blue and penetrating, and he always

Klaustur's memorial chapel to Jón Steingrímsson is built to evoke the A-shape of traditional Icelandic homes and barns.

seems to have just a slight grin around his mouth. Born and raised in the village, Helgason lived most of his professional life in Reykjavík, where he served in the country's parliament and as agricultural and justice minister. On retiring, though, he came back home as quickly as he could. Now he lives on a farm south of Klaustur, and spearheads efforts to preserve the town's history for future generations.

Helgason gestures out the window, through which we can see Klaustur's most famous building: a dark brown, modern A-frame wooden structure, just outside the old church grounds. The local women's association organised the fund-raising to get the chapel built, and a hundred farmers – true to tradition – each donated one autumnal lamb for six years to help pay for it. This is the Steingrímsson memorial chapel, consecrated in 1974 to commemorate Jón's stopping of the lava. It is also the spiritual and emotional centre of town: in 1983, a 200th-anniversary celebration of the Fire Mass brought several hundred people here in a moment of re-creation.

When Helgason takes us inside we see that the building is set up in simple Icelandic fashion, with wooden pews flanking the single aisle. It is a quiet and contemplative space, with views of the verdant cliffs above town. Off to one side of the pulpit sits a small architectural model under glass: a replica of the interior of the original church, where the Fire Mass was held. Its roof is authentically covered in miniature artificial grass, a nod to the days when almost every Icelandic building was A-frame in design and partly buried beneath the lush turf. A cut-out in the model reveals the simple wooden furniture within.

Looking up from the model, we see something above us and catch our breath. On each side of the chapel, attached to the A-frame, are two obviously ancient timbers. Splintery and fissured, the wood is smooth in places, scarred with black holes where nails used to be. Square-headed nails and rusty pieces of iron are embedded deep in the coarse grain. We look at Helgason and ask if these are what we think they are. And he nods. These are timbers from Jón's original church. Hesitantly, we touch them. These timbers are relics of the Fire Mass, the remnants of a now legendary battle between man and volcano.

The eerie feeling grows stronger as we leave the chapel and walk out to the grounds behind. Here, a low stone wall and a stand of birch trees enclose a wide squarish area. This is where Jón's church stood, where the Fire Mass took place. A simple white cross rises from the undulating ground, bearing a small plaque that reads 'Hér var Eldmessan 20. Júlí 1783' ('Here was the Fire Mass, 20 July 1783'). Helgason tells us that the cross stands where the entrance to Jón's church is believed to have been.

On the ground nearby, so low that we almost trip over it, is a low rectangular basaltic stone: it's the gravestone of Jón and Thórunn, at a spot that would have lain behind the choir of the church. Lichens and weathering obscure the Icelandic runes that are carved into it. The rest of the enclosure looks like a small park, except for several modern graves occupying the northern corner.

Around this small, carefully tended cemetery the ground rolls away, until it banks up to a mound in the southwestern corner. There are no markers here, for it is a mass grave. It's the burial site of some 76 parishioners who died during the Laki eruption and its aftermath. Yet another chill rolls down our spines: this is where Jón led his trusty horse, burdened with the corpses of his parishioners, and where he consecrated the bodies to the ground. So many died so quickly in the famine that they did not receive individual burials – though each, Jón was proud to say, was buried in his or her own coffin.

Here is indeed the true heart of Klaustur. It was sacred ground long before Jón arrived here. Just outside this stone enclosure, recent excavations have uncovered the remains of the town's medieval convent, whose nuns were celebrated for their skill in working textiles. (A loom has been unearthed here by archaeologists.) On old maps of Iceland, Klaustur is often designated with the sign of a cross, to indicate its famous convent. By Jón's time, however, the convent had fallen into ruin, and his church soon went the same way.

For nearly two decades after 1783, fierce winds blew volcanic sand from the lava deposits east of Klaustur, and by 1850 the grains had almost buried the building in a dune. Klaustur's congregation abandoned the church and moved its Sunday services a few kilometres north to Prestbakki, the former rectory for the parish and where Jón lived during the Laki eruption. A new church built of Danish timber was erected at Prestbakki in 1859, and still stands bright white above the lush meadowland. These days, the local priest doesn't have much call for his services, given the dwindling population. But once a month he still holds services at Prestbakki.

The memory of the Fire Mass runs deep even here. The altar at this newer church is a hand-carved pulpit depicting Jón, in all his neck-ruffed glory, preaching the Fire Mass. In the conical steeple, one of the bells retrieved from the original chapel – now oxidized with age – still tolls for services, weddings and funerals.

Standing outside this church at Prestbakki, Helgason describes the devastation of Laki, in a recitation that makes him

Jón Helgason keeps the memory of the Fire Mass alive.

sound more like an eyewitness than a story teller. He details what happened where and when: lava came up to a church in Skal on 12 June; on 15 June, it flowed below Skal and Holt but halted for a while; over in the lowlands, lava formed a lake; days later lava came within 500 metres of cultivated land; on 14 July people fled to the west. Helgason's memory is long, and it is as if a part of him lives in the eighteenth century. 'I can remember one-third of the time since the eruption,' he says matter-of-factly. He and other locals still divide history into the time before the fires and the time after the fires.

Like many people in and around Klaustur, Helgason traces his ancestry to a farmer who survived the eruption and the hardships that followed. Family lore holds that this relative, named Oddur, pulled through because his farm happened to be at a spot along a river that drained a different portion of the highlands. Even as Laki contaminated the Skaftá and other rivers with ash and lava, this stream ran with mostly fresh, untainted water. Other families starved for lack of food, but

Oddur could usually go to the stream and retrieve enough fish to feed his six daughters. Even today this place in the river is known as Oddsbúr, or 'Odd's pantry'.

Laki's ash still runs in Klaustur's veins. Talk for a while with Sveinn Jensson, the young manager of the Hotel Klaustur, and he will tell you about his trips to Switzerland and Cuba to get away from the village and explore the world. But press a little deeper and you'll learn that he was born and raised on the farm where, more than two centuries ago, Danish authorities making the rounds after the famine found an entire family dead in their beds, felled by starvation and malnutrition after trying to boil and eat their shoes.

Sveinn is a slim, dark-haired man with the unfailingly polite manner of someone trained in the hospitality industry. He worked his way up at the hotel, from cleaning rooms to working in the restaurant to becoming general manager by the age of thirty. So perhaps it's not surprising that he'd like to see Klaustur become a more prominent spot on tourists' itineraries. Laki would be the key to getting visitors into town. In Italy, Pompeii and Etna have built entire tourist industries around their volcanic histories, he notes. Why not Klaustur?

In 2011, the town got perhaps a bigger taste of volcanic tourism than it may have wished. That year, Klaustur endured an eruption that, while not nearly as bad as Laki, at least gave a taste of what those panicky days in 1783 must have been like.

Along Iceland's southern coast, 21 May 2011 was a peaceful spring day, sunny and still. Then evening came. Jón Helgason remembers looking up at about 5.30 p.m. and seeing what he thought was smoke rising above the hills behind Klaustur. He realised that another volcano must have gone off. Helgason had seen many eruptions in his eight decades, and so he wasn't worried. He went to check on his wife, who was in a nursing home not far from their farm, and told her everything was okay.

Some things never change: the Grímsvötn eruption of 2011 had farmers around Klaustur scrambling to get their livestock out of ashfall.

By the time he got back to his house, though, a dark mist was coming down from the mountain, heading straight for town. 'Then I was thinking, maybe it would not be okay,' he says. Soon the darkness descended, plunging the town into a midnight gloom that caught many people unaware. A shepherd, out in the fields when the murk came on, had to grope his way along the ground to find his way home. Helgason stayed inside, watching the ash blow and worrying about his wife.

Over at the hotel, Sveinn sprang into action. The hotel asked its guests to close all their windows, and sealed off air ducts to prevent ash from being sucked in. Across town, Trausti Ísleifsson was trapped in his family's summerhouse, worrying about the fate of his expensive quad bikes and vans.

The culprit for this eruption was the obvious one. Grímsvötn, the most active of all Icelandic volcanoes, lies under the Vatnajökull ice cap north and east of Klaustur. Ash from Grímsvötn regularly settles over Klaustur, and most townspeople regard the volcano as a background sort of annoyance, a part of the landscape as inescapable as rain or lightning. But this time Grímsvötn was not just a mild problem. The outburst of May 2011 would become the most powerful eruption in Iceland for half a century.

On the same evening that Helgason was watching dark clouds rise above Klaustur, geologist Bergrún Óladóttir was in Reykjavík getting ready for a party. It was the 50th birthday celebration for one of the University of Iceland's leading volcanologists, Magnús Tumi Gudmundsson, who has specialised in studying Grímsvötn. Just before everyone was scheduled to gather, word spread quickly that Grímsvötn was erupting. 'I was thinking, that's kind of good,' Bergrún says. 'Some people get fireworks for their birthday; others get an eruption.'

Bergrún had moved to Klaustur when she was an infant and her husband now works at the front desk of the Hotel Klaustur with Sveinn. Nowadays Bergrún divides her time between the town and the city, raising her children while working on her university studies. When Grímsvötn went off, she was in the right place at the right time: as a graduate student at the university, she was able to land a seat on one of the first flights that ferried scientists to study the volcano's ash plume. Within hours the plume had soared to an altitude of twenty kilometres, far greater than Grímsvötn's previous eruption in 2004.

From the plane Bergrún could see the ash billowing like the ash from Eyjafjallajökull the year before. 'You thought, oh my god is this happening again?' she remembers. 'I don't know how to describe it when you see things like this up close. It's just so much bigger than you'd think, and you can't do anything. It's just there, and you have to respect it.'

Her family was back in Klaustur, smack in the path of the oncoming black cloud. Was she scared? 'No,' Bergrún says. 'Just surprised at how big it was.'

Early the next morning, Bergrún and another geologist set out from Reykjavík, heading for Klaustur to collect ash samples. As they drove east on the ring road, Bergrún kept getting text messages from her father: the air was getting blacker; her parents could barely see outside the house.

Bergrún made it to Klaustur that afternoon, but when she and her colleague drove into town, the ashfall was too heavy to do anything but go inside and wait it out. Black debris rained down like an evil snowstorm. 'I lost all sense of direction,' she says. Anyone who ventured outside had to wear a mask, as well as sunglasses or goggles to keep the sand-like material out of their eyes. Winds blew dark gray ash through the town, where it piled up like a filthy snow on the roofs.

Many people evacuated. Trausti and his brother gave up after three days and drove back to Reykjavík, as did Sveinn's parents. For those who stayed, emergency officials and the local priest telephoned every house in town to make sure everyone was all right. One saving grace was that communications remained intact. Electricity stayed on, and phone and Internet connections held, so people could stay abreast of the news even though they couldn't see anything out of their windows. This part was crucial to maintaining calm and order: everyone had access to information and knew what was going on.

In the end, the 2011 eruption of Grímsvötn spewed out twice as much material as Eyjafjallajökull had done the year before, and in one-tenth the time. Ash blackened skies around the entire island, but because winds weren't blowing toward Europe, most planes kept flying as planned. Some flights were cancelled, notably in Scotland and Germany, along with those from Reykjavík, but Heathrow airport remained open. Which is why most people neither heard nor cared about Grímsvötn. This time, Iceland kept its volcanoes close.

After bidding goodbye to Klaustur, we wend our way back to Reykjavík to spend a few more hours in 1783 Iceland: we're

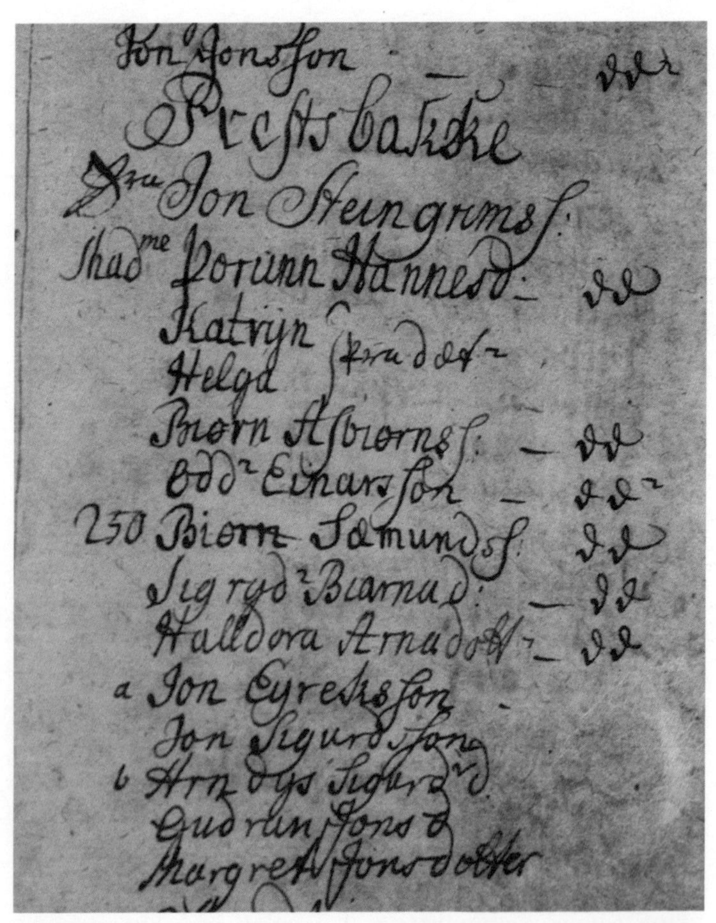

Jon Jonsson — dd
Prestsbakke
Sra Jon Steingrims s.
Madme Þorunn Hannesd: — dd
Katrijn } [illegible]
Helga
Biorn Asbiorns s. — dd
Oddr Einarsson — dd
250 Biorn Sæmunds s. dd
Sigrydr Biarnad: — dd
Halldora Arnadott — dd
a Jon Eyreksson
Jon Sigurdsson
b Arndys Sigurd d.
Gudrun Jons d
Margret Jons dotter

A page from Jón's Eldrit, or Book of Fire, records the members of his household at Prestbakki (and others). Below his name is that of his wife, Thórunn Hannesdóttir, marked like others with two squiggles that denote their deaths.

hoping to get a look at the manuscript of Jón's chronicle of the eruption. It includes, in his words, 'all that I had written down, day by day and interval by interval, while the fiery chastisement lasted until God gave relief and brought this plague on our country to a fortunate end.' His stated purpose in writing it was to serve as a warning to his fellow Icelanders so they could learn about 'the castigation of the Lord . . . for their own betterment.' But it has endured as a scientific document, and is now preserved for the country's heritage at the National and University Library of Iceland in Reykjavík.

The library itself is an impressively modern three-storey complex, with the main façade painted a rich red, the same colour as the roofs of traditional Icelandic farms. We make our way past exhibits on Víking history and descend into the heart of the building. Jón's text is stored in the manuscript collection, which holds some 15,000 items, dating back to the year 1100. We put in our request with an elderly gentleman who nods and motions us toward a pile of white archivist gloves on a lower shelf. He disappears through a pair of double doors, and we wait.

After a while he emerges holding a small parcel, a four-flapped envelope tied together with laces. He says nothing, but places it on the table. The title on the envelope's cover reads simply 'Eldrit of Reverend Jón Steingrímsson.'

On go the white gloves. We undo the laces, and the envelope flaps open to reveal the manuscript cover. It is in the style of a composition notebook, except that the cover's marbling is a sombre blood-red. It's smaller than we thought it would be, about six by eight inches. Gingerly we open the cover and begin to page through this remarkable document. For a manuscript written in the eighteenth century, it is in excellent condition, and the meticulous, calligraphic script is still quite legible. We can imagine Jón painstakingly bringing his exacting penmanship to bear in the dark of night, a candle or fish-liver-oil lamp providing the only illumination.

We've read the main text many times in English translation, so we can pick out a few relevant Icelandic words in the first part of the document. But then we come to the back pages, where Jón has appended additional material not in the English version. It contains a census listing all the people who died in the various parishes and farms in Jón's district. We recognize the names of villages as well as some personal names. Under the listing for the farm at Prestbakki, we see Jón Steingrímsson's name and, below that, his wife Thórunn Hannesdóttir. Beside her name are two curious squiggles – a hieroglyph that Jón uses throughout the census to signify that a person is deceased. What, we wonder, must have gone through his mind as he officially inscribed the death of his own wife?

Five other names in Jón's household are also noted as deceased – three men and two women, most likely workers on his farm. In all, some 224 names in this census bear Jón's mark of death. Some would have been family; all would have been friends.

The last passage of the text contains two poignant notes. The first is a solemn report that the dead 'were buried in the graveyard behind the church'. The second footnote states Jón's wish to be buried 'just to the side of my dear wife Thórunn where I also want to be buried within sight of her if God allows me to die here (as I hope).'

Jón got his wish.

CHAPTER EIGHT

Death by Volcano

The many ways eruptions can kill

THE KILLER ARRIVED IN THE QUIET CAMEROON EVENING, as people were sitting around fires talking or making their way to bed. Those who survived say the horror came swiftly. Around 9 p.m., on 21 August 1986, they heard strange explosions and the sound of a strong wind. Then came the smell, like rotten eggs. Almost immediately people lost consciousness, only to wake hours later with blisters on their skin and red marks on their clothing. Around them lay the corpses of people, cattle and other animals. Some malicious spirit had raced silently through the village, leaving devastation in its wake.

Months later, survivor Joseph Nkwain told his story of that night:

> *My skin became very hot and I perceived something making some dry smell. I could not speak, I became unconscious, I could not open my mouth because then I smelt something terrible and could not speak. I just closed my mouth and remained silent. All of a sudden, I heard my daughter snoring in a terrible way, very abnormal. So I forced myself to stand up from the bed, I was already weak . . . When*

crossing to my daughter's bed, in the middle of the floor, I collapsed and fell ... I was there until a friend of mine came and knocked at my door. The door was locked, he hit it very loudly, so much noise that he woke me. I heard it as if I was dreaming, I was surprised to see that my trousers were red, had some stains like honey: I saw some starch, some starchy mass on my body. My arms had some wounds ... My daughter was already dead.

Nkwain and his family lived on the shores of Lake Nyos, a body of water nestled in Cameroon's northwestern highlands. In one quick exhalation, the lake had released a deadly cloud of volcanic carbon dioxide, which slithered down the valley bottoms and suffocated at least 1,700 people in a single night.

The lake isn't obviously linked to any active volcano, and the nearest recent eruptions are 300 kilometres away. But Lake Nyos fills a crater that is part of the Cameroonian volcanic chain, a line formed where two parts of the African continent are pulling apart from one another. Even though fresh lava isn't erupting on the surface, magma seethes not too far below. That underground heat, along with volcanic gases, feeds the well-known soda springs at lakes such as Nyos and its neighbour, Monoun.

Monoun had struck two years earlier. On 15 August 1984, a gas cloud erupted from that lake and smothered 37 people. The news hadn't spread very far, in part because the Cameroon government tried to keep things quiet: when they found bodies marked with burns and blisters, they suspected chemical terrorism, so they hushed it up. After the Nyos disaster, however, headlines erupted. Scientists and humanitarian groups descended on the area, collecting stories from survivors and trying to analyze what had happened. It was an exercise in deduction worthy of Sherlock Holmes. Bodies were found as far as 25 kilometres from the lake, in a pattern that hinted at the spread of some toxin. Many corpses remained inside houses. Some people had died instantly; others had managed

A 200-metre-long pipe helped degas the deadly carbon dioxide lurking in the depths of Lake Nyos, Cameroon.

to grasp torches and flee into the bush, only to be struck down by the invisible killer. Those who survived usually had no memory of what had happened, other than a vague sense of confusion followed by unconsciousness.

Plants by the lake shore were flattened in places, as if something had rolled over the reeds. But they showed no signs of scarring or other damage that might be expected if the cloud were extremely hot or filled with sulphuric acid. That's why scientists eventually converged on the carbon dioxide explanation: at carbon dioxide concentrations higher than about 10 per cent, people become confused; at concentrations higher than about 30 per cent, people pass out into a reversible coma – or die.

But where had all the gas come from? Researchers eventually figured out that it had come from the bottom of Lake Nyos. Volcanic springs percolating into the lake bottom had injected carbon dioxide into the lower layers of the lake's 208-metre depths. An inversion layer closer to the surface formed a sort of lid, trapping the dissolved gas at depth. Pressure built up until suddenly, disturbed perhaps by an underwater landslide, the trapping layer was breached and gas-rich waters erupted from the lake's depths. It was like taking the cork off a bottle of champagne: bubbles burst forth, strewing carbon dioxide everywhere. And that gas, which is about 1.5 times as dense as air, flowed serpent-like along the ground, suffocating any living thing in its path.

Just as frightening, the 1986 disaster didn't seem to have emptied Nyos of all the deadly gas lurking at depth. Scientists who visited the lake after the eruption lowered sampling bottles into Nyos; when they brought them up, the bottles exploded. Plenty of carbon dioxide was still dissolved there, a killer in waiting. In 2001 a French-Cameroonian team decided to see if it could de-gas Nyos by lowering several large pipes vertically into the lake and allowing water to shoot out like a Versailles fountain gone berserk. Ten years later, most of the pressurized gas had been vented off, and at nearby Monoun, three pipes inserted between 2003 and 2006 had completely de-gassed the lake by 2009. (The scientists are now working to see if they can use the same idea to extract methane for energy development from Lake Kivu, on the border between Rwanda and the Democratic Republic of Congo.)

Science may have tamed Africa's killer lakes for now, but local memory runs deep. The late anthropologist Eugenie Shanklin argued that people living in this part of Cameroon had experienced deadly lake outbursts before, and that oral histories preserve a record of these disasters. At least thirty ethnic groups live in the Cameroonian highlands, including a group known as the Kom people. Their complex origin story includes a tale of a battle between a Kom leader and his counterpart from another group. On being outwitted, the Kom leader hanged himself but ordered that his body should not be cut down. His bodily fluids instead dripped to form a lake, and maggots from his corpse populated the water as fish. The other leader brought his group to this miraculous new lake – whereupon the lake exploded and sank into the Earth, taking all the Kom's rivals with it.

Another ethnic group, at the crater lake of Oku in Cameroon, has a similar story in which the waters of the lake leave its bed and destroy the homes of rivals. 'Oku people still make elaborate annual sacrifices to the lake, which showed by its actions that it wished to belong to the Oku people,' Shanklin wrote in 1989.

Perhaps because of such tales, most Cameroonians live several miles away from these dangerous crater lakes. It turns out that at Nyos, the only people living by the water were a group who arrived in the 1940s and 1950s. They were apparently unaware that lands by the shore were unoccupied for a reason.

Lake Nyos and its lethal clouds are far from isolated cases. There are an impressive number of ways in which volcanoes can kill you, and history records more than a quarter of a million deaths caused directly by some four hundred eruptions. When you take into account the many indirect deaths from volcanic eruptions, that number rises far higher.

For a start, you can be incinerated in a fast-moving lava flow. This type of death is a favourite of Hollywood scriptwriters, but it doesn't happen very often in real life. When it does, though, the consequences can be dire: in 1977, a lava lake at Nyiragongo volcano in Zaire drained suddenly and unpredictably. Its lava was so hot and its chemistry so unusual that it flowed very quickly, swallowing villages and killing hundreds if not thousands of people.

Far more dangerous than most lava flows are the sudden surges of hot gas, ash, rock fragments and other debris known as pyroclastic flows. The word comes from the Greek pyro (fire) and klastos (broken), reflecting how a volcanic explosion can blow magma and rock into tiny pieces. Pyroclastic flows can happen if an eruption ejects material that is heavier than the surrounding air: it collapses into a fluid-like avalanche that might move at speeds of more than 100 metres per second. They are what obliterated Herculaneum during the 79 C.E. eruption of Vesuvius, and Saint-Pierre during the 1902 eruption on Martinique, which knocked over metre-thick stone walls. If you're buried by a pyroclastic flow, chances are that your body will never be found.

Even modern scientists, with the best training and best equipment, can be caught unawares. French photographers Maurice and Katia Krafft, who between them had probably observed more eruptions than most twentieth-century scientists, perished in June 1991 at Unzen, Japan, when a pyroclastic flow stopped following the path that others before it had taken, and instead swept up and over the ridge on which they were standing. Some forty people died in Unzen's eruption that day, including American volcanologist Harry Glicken. (Glicken had cheated death before. He had been observing Mount St. Helens for the U.S. Geological Survey in May 1980 when he left for a college interview; his replacement on the job, David Johnston, was killed in the blast.)

Even a relatively small pyroclastic flow can have devastating consequences, since they behave so unpredictably. Firebrands can detach from the main current and start secondary fires or currents nearby. And flows that are relatively dilute – that is, not carrying so much debris – can rush up steep slopes, or change direction quickly. It's nearly impossible to tell just by looking at a pyroclastic flow which kind you might be facing.

Then there's the danger of ash and other rock debris (known collectively as 'tephra'), which can overload the roofs of buildings, causing them to collapse. In many volcanic areas, buildings are reinforced to lower the risk, and people are educated to sweep their roofs clear of ash, or to evacuate if the ashfall gets too heavy to keep up with. Out in the open, you're even more vulnerable. At Mount St. Helens, eighteen people were engulfed by tephra and asphyxiated by plugs of ash and mucus.

Another risk from eruptions is the presence of large rocks moving at near-ballistic speeds. During the Mount St. Helens eruption, a man died from a rock that smashed into his car 16 kilometres from the main volcanic vent. Scientists, too, are not immune: at least twice on field trips after international volcanology meetings, researchers have been killed on visits to observe local eruptions. In Colombia in 1993, six scientists and three others died at Galeras volcano when a sudden explosion occurred as they were standing inside the inner crater. In Indonesia in 2000, two volcanologists died and five others were injured in much the same circumstances.

For sheer unpredictability, though, almost nothing matches what can happen when an eruption creates a flood of water, mud and other material. Known as lahars or debris flows, these torrents often have the consistency of wet concrete, and can destroy everything in their path. Many of the world's most active volcanoes are high mountains capped with snow and ice: when these erupt, the heat melts the ice and generates deadly lahars. In 1985, nearly the entire city of Armero, Colombia, was annihilated by mudflows coming from the eruption of the glacier-topped Nevado del Ruiz.

Icelanders, of course, suffer glacial outburst floods as well, the jökulhlaups. The 1996 eruption at Grímsvötn drained a subglacial lake, sending a torrent rushing along a 50-kilometre path. The biggest flood ever measured at Grímsvötn, with some 40,000 cubic metres pouring out every second, it washed away a portion of Iceland's crucial ring road and destroyed two bridges, including one nearly a kilometre long.

Floods are not the only watery disasters linked to volcanoes. Under the right conditions, volcanoes have the potential for producing catastrophic tsunamis. Usually the sheer force of the explosion triggers underwater landslides that kick up massive

Multiple pyroclastic flows left their deadly traces down Unzen volcano, Japan, seen here in November 1991.

waves that then race ashore. In 1792, a tsunami precipitated by a landslide during an eruption of Unzen may have washed away more than 10,000 people. After the 1815 eruption of Tambora, pyroclastic flows rushing into the sea generated waves more than ten metres high that killed at least 4,600. And most of the 36,000 deaths at Krakatau in 1883 probably came from a tsunami created when the volcano collapsed suddenly into the ocean.

Finally, there are the toxic gases. Killers such as carbon dioxide and hydrogen sulphide suffocate anyone unlucky enough to run into them, as at Lake Nyos. But volcanic gases don't always kill right away. Prolonged exposure to them can cause severe health problems in people living near volcanoes.

One of the best places to study such health hazards is at Kilauea, Hawaii, which has been erupting continuously since 1983. Perched on the southeast corner of what Hawaiians call simply the Big Island, Kilauea is one of the world's best-studied volcanoes. MIT professor Thomas Jaggar started it all in 1912, when he employed prisoners from a nearby military camp to dig through six feet of pumice and ash to install seismometers on Kilauea's steep caldera rim. This single observation post would later morph into the Hawaiian Volcano Observatory, the first long-term scientific effort of its kind.

Jaggar was moved to act after 8 May 1902, the day when 30,000 people died in the eruption on Martinique. Hours earlier, 1,700 had perished on the nearby island of St. Vincent, in an eruption of La Soufrière. Jaggar was part of an expedition of scientists sent to document the damage from both Caribbean volcanoes. Horrified by what he saw, Jaggar was particularly upset by the fact that Martinique officials had made no efforts to warn people of the volcano's danger. He thought scientists should do better.

Thus was born the idea of a permanent observatory, to better understand the behaviour of volcanoes and develop ways of predicting eruptions. 'If we could get . . . a properly endowed laboratory of the study of earth movements . . . we might be able in a few years to make earthquakes and volcanoes ordinary risks for insurance, and also succeed in preserving a great many human lives,' Jaggar wrote in a San Francisco newspaper in June 1906, two months after an earthquake devastated that city.

Jaggar spent a long time thinking about where he wanted to set up his observatory, and visited Japan, Alaska, the Caribbean and Italy before settling on Kilauea, which had the advantage of being on American territory and fairly easy to reach. Plus, he wrote, the Hawaiian volcanoes 'are famous in the history of science for their remarkably liquid lavas and nearly continuous activity.' By January 1912 the observation post was up and running.

Today the Thomas A. Jaggar Museum and the Hawaiian Volcano Observatory perch together on Kilauea's summit. This is very near the spot where Kilauea erupted explosively in 1790, in the deadliest eruption on any territory that would become part of the United States. Some one hundred warriors and their families were killed as they walked past the crater, during a great battle with another local chieftain. The thousands of visitors who stream through the museum and visitors' centre every year are walking practically on top of this deadly spot.

The museum conveniently overlooks a large crater, from which steam rises ominously and a bright red-orange glow can be seen at night. This crater, known as Halemaumau, is the legendary home of the Hawaiian fire goddess Pele. Her older sister, Namaka-o-kahai, was a goddess of the sea, and the two siblings fought constantly. Namaka-o-kahai eventually ran Pele out of their home, and the banished sister has since languished alone at the smoking summit of Kilauea. When Pele gets angry, she spits lava and hisses steam. Tradition has it that the only way to placate her is to leave small offerings, like the leis and other tokens often found dotting the trails around Kilauea.

Since 1983 Pele has taken out her wrath at various spots along Kilauea's huge, dome-shaped surface. In addition to the summit crater of Halemaumau, much of the action takes place along an eastern rift zone known as Pu'u O'o. Photographers congregate to capture glowing red lava plunging dramatically into the sea with much hissing and steaming. Other portions of the rift zone have also opened up over time, spilling lava into established communities and swallowing whole townships. Not that this keeps all Hawaiians away: brand-new homes have already cropped up on lava flows that stopped steaming just a few years ago.

With people and eruptions so close together, it's no wonder that Kilauea has become a favoured place for studying the

Pele, the Hawaiian fire goddess. Note the rope-like lava that constitutes her hair.

health hazards of volcanoes. Hawaiians are familiar with the thick haze they call vog, or volcanic smog. When trade winds blow from the northeast, the vog often wraps around the Big Island from Kilauea and settles over the more densely populated western coast. When winds are variable or blow from the south, vog can spread all the way up the Hawaiian island chain, to Oahu (site of the capital, Honolulu) and beyond.

Near the active vents, vog is made mostly of sulphur dioxide droplets. Over time and at greater distances, those droplets react with sunlight and chemicals in the air to become more complex sulphur compounds, including sulphuric acid. Breathing in this stuff isn't great for you: sulphur dioxide irritates the throat and nose, and aerosol particles can lodge in the lungs, aggravating asthma.

Bernadette Longo, an epidemiologist at the University of Nevada in Reno, has looked at rates of asthma and other respiratory illnesses in people living on the Big Island, and found major hazards coming from Kilauea. Things got particularly bad after 2008, when the summit crater of Halemaumau began to erupt, disgorging about three times as much sulphur dioxide as before. Levels soared well above the World Health Organization's 24-hour guidelines. Between 2004 and 2010, Longo found, the risk of respiratory illness went up dramatically for anyone living in vog-thick areas: asthma went up 222 per cent (579 per cent for children); acute bronchitis increased 73 per cent (444 per cent for children); and upper respiratory infections 83 per cent (234 per cent for children).

Longo's findings from Hawaii may be particularly relevant to Laki, and the many months during which Icelanders sucked its toxic gases into their lungs.

After all, many of the deadliest volcanic phenomena were not a problem at Laki. There were no pyroclastic flows, killer jökulhlaups or powerful tsunamis. The eruption occurred relatively far from human settlement, and people in and around Klaustur could generally get out of the way of lava flows in time.

Yet census figures show that Iceland's 1783 population of 48,884 was reduced three years later to 38,363 – a loss of 22 per cent. Officially these deaths all resulted from famine. However, starvation may not explain everything. In particular, one underappreciated culprit may be the element fluorine.

Sheep, horses and other mainstays of the Icelandic economy are vulnerable to a fluorine poisoning known as fluorosis. Among its earliest symptoms are the famous pointy 'ash-teeth' first noted after a seventeenth-century Hekla eruption. Icelandic farmers today still occasionally put bowls of water outside, to see if they catch any ash particles falling from a nearby or distant eruption; if any ash is detected, the farmers move the animals inside. Fluorine concentrations are actually highest in ash particles farthest from the eruption, since most of the element is carried by the finest-grained particles.

More so than many eruptions, Laki was particularly rich in fluorine. Maureen Feineman, a geochemist at Pennsylvania State University, has studied the chemical composition of rocks formed during the 1783–84 eruption. Her work found that the magma that eventually erupted from the Laki fissure must have rested underground for some time, melting and mixing in some of the rocks surrounding the magma reservoir. That extra time sitting in the crust led to the absorption of increased levels of elements such as sulphur, chlorine and fluorine. With more of these volatile elements to start with, Laki was able to spit large amounts of them out.

Fluorine poisoning occurs in people living near other volcanoes. In the Azores, people have contracted fluorosis from drinking water that circulated through fluorine-rich ash beds. In Vanuatu, where people collect drinking water on their roofs during rainstorms, up to 96 per cent of children living near the constantly de-gassing volcano of Ambrym display signs of dental fluorosis. In Iceland, too, fluorine would have settled on the ground and run into rivers from which villagers took their drinking water. At its peak, the 2010 Eyjafjallajökull eruption was showering the landscape with nearly 700 tonnes of fluorine daily.

Even one of Iceland's greatest heroes might have suffered from fluorosis. Egil, son of Skalla-Grím, who died around the year 990, is the protagonist of a thirteenth-century saga detailing his raids with Vikings across much of northern Europe. For all his prowess in battle, the saga tells that Egil was a funny-looking and often sick man. His bones were 'much bigger than ordinary human bones', and his skull was 'an exceptionally large one and its weight was even more remarkable. It was ridged all over like a scallop shell.' All the better, perhaps, to weather massive blows from the axe of another Viking.

The saga hero Egil, possibly a victim of volcanic fluorine poisoning.

Whoever wrote Egil's saga took care to include these detailed descriptions of his bone deformities. Among other ailments, the warrior suffered from deafness, lethargy, headaches, cold feet and occasional blindness. Some scholars have attributed Egil's problems to Paget's disease, in which bones become enlarged and misshapen, particularly around the face. According to this theory, Paget's disease could account for why Egil described himself as having 'a helm's-rock of a head'.

But Philip Weinstein, a paleopathologist at the University of Western Australia, thinks fluorine could be the explanation. The way Egil is described in the Icelandic sagas is consistent with his suffering from skeletal fluorosis, Weinstein argued in a 2005 paper. Symptoms such cold feet and blindness aren't easily explained by Paget's, and whereas people suffering from Paget's often break their bones, Egil's bones seem to have been rather robust: more than a century after his death, his skull was reportedly dug up and found to be capable of withstanding a whack from an axe.

Returning to Laki, we see that Jón Steingrímsson's chronicle describes symptoms that sound very much like fluorosis. So too does the official report from the Danish-appointed chronicler Magnús Stephensen:

> *The same symptoms shewed [sic] themselves, in this disorder, in the human race, as among the cattle. The feet, thighs, hips, arms, throat and head were most dreadfully swelled, especially about the ankles, the knees, and the various joints, which last, as well as the ribs, were contracted. The sinews, too, were drawn up, with painful cramps, so that the wretched sufferers became crooked, and had an appearance the most pitiable. In addition to this, they were oppressed with pains across the breast and loins; their teeth became loose, and were covered with the swollen gums, which at length mortified, and fell off in large pieces of a black or sometimes dark blue colour. Disgusting sores were formed in the palate and throat, and not uncommonly at the extremity of the disease, the tongue rotted entirely out of the mouth. This, dreadful, though, apparently, not very infectious, distemper, prevailed in almost every farm in the vicinity of the fire during the winter and spring.*

In 2004, medical geologist Peter Baxter, of the University of Cambridge, worked with Icelandic colleagues to exhume three bodies buried in cemeteries near Laki. Their goal was to search for signs of fluorine poisoning in the skeletons, such as deformed bones and teeth. Unfortunately, two of the bodies turned out to be at least a half-century too late to have died during the Laki eruption, and the third skeleton showed no evidence of fluorine overdose. For now, the jury is out.

Tens of thousands – possibly hundreds of thousands – perished across Europe in the aftermath of Laki, and they did not die from fluorine poisoning. Instead, these victims probably succumbed to a combination of factors.

In 2005, two British volcanologists ventured into historical demography to try to work out how many people Laki may have killed in England. Cambridge scientists Claire Witham and Clive Oppenheimer studied an exhaustive database that included the monthly and annual frequencies of baptisms, burials and marriages in the registers of 404 parishes in 39 counties between 1538 and 1871. The data aren't perfect: the records cover only a small percentage of the country's population, are not evenly distributed among the counties, and do not state the cause of death. Still, they are sufficient to reveal out-of-the-ordinary trends.

Witham and Oppenheimer discovered that the late summer burial rate in 1783 was the highest recorded in the entire eighteenth century. They also found two spikes in death rates: between August and September 1783, mortality was 40 per cent higher than the mean, and between January and February 1784, death rates were 23 per cent higher. Together, these two events accounted for about 20,000 extra deaths across England.

What could explain these deaths? It's likely that many were linked to the exhaustion caused by the sweltering heat of the summer of 1783. Insects might also have played a role, in spreading food contamination or disease. Witham and Oppenheimer argue that the warmer temperatures may have bolstered the transmission of diseases that have longer life cycles, such as malaria, which is transmitted by a parasite in female *Anopheles* mosquitoes and which takes several weeks to progress from mosquito infection to human infection. The seventeenth-century English physician Thomas Sydenham described the fevers, or agues, that he associated with the appearance of insects: 'When insects do swarm extraordinarily and when . . . agues appear as early as about midsummer, then autumn proves very sickly.' The time-lag of malaria infection could explain why the summer death toll spiked in August and September 1783 rather than in June and July, when temperatures were hottest. Another possibility is diseases like typhoid and dysentery, which spread more readily during heat waves but take longer to kill. Remember Gilbert White's observations of swarming flies and spoiling meat; this sort of contamination could have dire consequences.

Winter brought a whole new suite of health problems. As Laki's aerosols chilled the continent, people were forced to huddle together for warmth. Diseases such as typhus, which is carried by lice, may have spread more readily. Anyone already weakened by the difficult summer could have then been finished off by the cold and disease of the following winter. This could explain the second mortality spike, in January and February 1784.

Finally, the dry fog itself may have played a major role in England's mortality spikes. Bernadette Longo has shown the effects of breathing Hawaiian vog over many months, and the same thing was happening across Europe in 1783. Anywhere the volcanic haze persisted, people were breathing it in – with all the attendant consequences for their health.

There are plenty of modern examples of how pollution can kill. In October 1948, an atmospheric inversion trapped industrial smog over the small mill town of Donora, Pennsylvania. For five days an acrid haze blanketed the valley. Residents tried to make the best of it, even sending their children out to march as usual in the annual Halloween parade. But within days people were crowding doctors' surgeries and hospital rooms, complaining of choking and burning in their eyes and throats. The smog had appeared on a Wednesday; the first person died around 2 a.m. the following Saturday. By the time fresh air blew in and cleared out the haze, on Sunday afternoon, twenty people had died. The Donora smog was one of the worst environmental disasters in US history, and served as a trigger for state and federal regulations and eventually the revolutionary Clean Air Act of 1970.

Donora's noxious smog is not an exact analogue for Laki's volcanic haze, but it does show the kind of health hazards people face when breathing in a heavy load of particulates for days on end. So too does another great pollution disaster, one which may be more directly relevant to the Laki experience. In the winter of 1952, London experienced its own Donora moment – the so-called Great Smog. On 5 December, a thick haze descended over the city, where weather patterns trapped it for the next four days. But unlike Pennsylvania, where the steel plants did the damage, London was doomed by people simply going about their everyday business.

Early December 1952 was bitterly cold, and Londoners had been burning more coal than normal. The low-grade domestic coal used at the time released copious amounts of sulphur dioxide in the smoke. Air pollution measurements made by the London County Council found concentrations of sulphur dioxide as high as 3,830 micrograms per cubic metre. The World Health Organization defines the safety threshold for sulphur dioxide exposure as 20 micrograms per cubic metre for an entire day, and 500 micrograms for a ten-minute peak.

Yet in December 1952 people continued to go outside, covering their noses and mouths with masks or handkerchiefs. The streets were so dark that cars had their headlights on at noon. A performance of *La Traviata* began on time at Sadler's Wells, only to be halted after the first act because of smog inside the theatre.

Within days people began to die. They would continue to die for months to come, long after the smog had dissipated. The weekly number of extra deaths in Greater London (the number of deaths exceeding those recorded during the same time period the previous year) peaked at about 4,500 for the week ending 13 December 1952. Total mortality rates for

The Great London Smog of December 1952 had residents wearing face masks to protect themselves from the sulphureous pollution.

the month were an astonishing 80 per cent higher than the previous year, and 50 and 40 per cent higher for January and February 1953.

A 1954 report by the UK Ministry of Health suggested that most of these people may have died not from air pollution, but from influenza. But a 2004 analysis reviewed flu reports from the time and calculated that only a fraction of the excess deaths could be explained in this way. Also, analyses of lung tissue from people who died during the Great Smog found soot and other particle types in the lungs. And a review of autopsy records at the Royal London Hospital revealed that deaths from chronic obstructive pulmonary disease between December 1952 and February 1953 were double those of corresponding months in other years.

The haze in the summer and autumn of 1783 would have been far, far worse than the Great Smog of 1952. Whereas the London fog persisted for only four days, the Laki fog enshrouded the continent almost continually between June and September or October – even, in some cases, December. For months, Laki disgorged millions of tonnes of acid aerosols daily. Sulphur dioxide concentrations across Europe would have surely passed critical thresholds for human health time and again.

Reports from the time are eerily similar to those of modern pollution events. The stench of sulphur hung in the air, and people struggled to breathe. In Champseru, France, a 'pestilence' of the throat lingered for ten months, long after the Laki fog had dispersed. In the Netherlands, Sebaldi Brugmans reported that people suffering from asthma had a particularly hard time during the Laki haze. Environmental historian John Grattan has suggested that this may have been the first time anyone explicitly linked bad air quality to worsening asthma symptoms.

Most of the respiratory symptoms described across Europe in 1783 can be explained by the sulphur, chlorine and fluorine that Laki produced. And it's more than possible that Laki could again kill hundreds of thousands across Europe.

Anja Schmidt is a rising star in volcanology. A slight and intense woman, she grew up in East Germany and began her career in information technology, but soon saw that opportunities would be greater – and perhaps more fun – if she returned to university and studied volcanoes instead. Today she is an atmospheric modeller at the University of Leeds, using computer simulations to test how pollutants and other materials move around in the atmosphere.

Sitting at a picnic table outside one of Iceland's famous hot dog stands, Schmidt tells us how she became interested in Laki. She had been awarded a student fellowship at Leeds that would let her pursue any topic of interest. She wanted to use the university's climate model to study how volcanic eruptions might affect the atmosphere. For that, she needed data on historical eruptions, and the eruption of Laki – whose emissions had been painstakingly calculated by scientists such as Thor Thordarson and Steve Self – provided exactly what she required.

'I don't want to do a study with no implications for society,' Schmidt says. So she tackled the question of what might happen if a Laki-like eruption were to go off tomorrow. Give the Leeds computer model a certain number of factors – how much material the volcano ejects, for how long, and at what altitude – and it can create a prediction of where that material might spread. For Laki, Schmidt programmed the model to erupt the same sort of stuff that Laki produced in 1783–84, but under today's atmospheric conditions. Her goal: to see how air pollution from such an eruption would spread across Europe today.

Schmidt focused on the spread of particles smaller than 2.5 micrometres. This size, known as 'PM 2.5', is a standard epidemiological measure for particle sizes that cause respiratory distress. High concentrations of particles smaller than PM 2.5 can lead to breathing problems, especially in children and the elderly. This is true whether the particles are made of sulphur dioxide, soot, or any other material.

Schmidt's paper, titled 'Excess mortality in Europe following a future Laki-style eruption', makes for depressing reading. During the first three months of the eruption, volcanic particles would cause air pollution loads in central, western and northern Europe to double. For 36 extra days the PM 2.5 air quality standard would be above the recommended guidelines from the World Health Organization. As a result, Schmidt's team calculated, some 142,000 people would die. Take another look at that number: 142,000 deaths.

Surprisingly, her work received relatively little press when it was published in the September 2011 issue of the Proceedings of the National Academy of Sciences. Perhaps people were inured after the drumbeat of coverage of the 2010 Eyjafjallajökull eruption. Or perhaps the devastation from another Laki just seemed a little too hypothetical. In January 2012, however, the UK's National Risk Register for the first time listed volcanic eruptions as something the country needs to prepare for. The list cited Schmidt's study.

Still, the question remains: what could governments actually do if such an eruption occurred?

CHAPTER NINE

The Next Big Bang

How worried should we be?

Nearly two decades before the mountain blew, it quivered. First came the earthquakes – small ones of magnitudes one, two and three, detectable by seismometers but not anything that farmers going about their business in Eyjafjallajökull's shadow would have noticed. Then the volcano began to move. Global-positioning stations resting on its flanks lifted ever so slightly higher, as if it were drawing in a deep breath.

After years of grumbling and spitting, the volcano began to awaken for real. In late March and April 2009, quakes began rattling beneath Eyjafjallajökull, as magma moved upward in its belly. Seismologists listened as the tremors became bigger and more frequent. By March of 2010 the volcano was shaking loudly. Its eastern flank began bulging upward as magma gathered beneath it.

Eyjafjallajökull finally roared to life on 20 March. Fire fountains exploded out of a rocky ridge just east of the mountain's ice-covered summit. It was one of the prettiest eruptions in years. During the day, the dusky light of a northern spring noon highlighted the spectacular orange bursts, while at night the cool green glow of the northern lights shimmered overhead. The eruption, on a flank known as Fimmvörduháls (Five Cairns Pass), was happening on a popular trekking path. Hikers now hired jeeps or quad bikes to run them up the icy trails and drop them within metres of the eruption.

Icelanders call this sort of event a 'tourist eruption', and enterprising operators quickly started running daily trips from Reykjavík. Vehicles were soon jockeying for position up and down the icy ridge, while police tried to corral the crowds into strict safety zones. It didn't always work: two people died of exposure after they drove too far from the eruption and their car ran out of fuel.

Had the volcano stopped at that point, Eyjafjallajökull would have been not much more than a footnote in the history books. But on 12 April lava ceased fountaining out of the barren ridge of Fimmvörduháls. A day and a half later, everything went to hell.

At 10.29 p.m. on 13 April, quakes started shaking directly underneath the snow-capped summit of Eyjafjallajökull, just to the west of the tourist eruption. For two and a half hours they went off like gunshots. At one point, as magma gushed upward, quakes were coming nearly once a minute in the five kilometres of rock directly below the summit. At around 1.15 a.m. on 14 April, the quakes started slowing down and a low-frequency tremor began reverberating through the top of the mountain. Icelandic seismologists later concluded that at this point the magma had emerged and begun melting the mountain's ice cap. Every second, fire from within the Earth was melting some 300 to 500 cubic metres of ice.

Because of the risk of meltwater floods, officials began evacuating residents from the south side of the volcano, and then the east and north. The first jökulhlaup rushed down the north side of the mountain in the early morning of 14 April. Today you can see the traces of its fury if you take a jaw-rattling ride up the valley of the Markarfljót river. Your driver will barrel past tranquil farmhouses, through icy streams, and across the great grey valley floor before coming to a halt on Eyjafjallajökull's north flank. The mountain looms above you, dark and ominous, its ice-capped peak usually enshrouded in white clouds. Descending towards you is a massive finger of ice, the glacier known as Gigjökull. The ice is deep blue at its heart, white closer to the surface, and dusted everywhere with the grey-black ash of the eruption.

Before April 2010, this frozen hulk reached down the side of Eyjafjallajökull and plunged into a beautiful glacial lagoon, where icebergs studded the jewel-blue water. All that changed in an instant when the volcano erupted and meltwater began roaring down. The flood coursed down the mountainside and spilled into the river valley so powerfully that it washed away the tranquil lagoon. Even the gauge put there to measure such an outburst was obliterated as a slurry of ice, rock and water rushed down the Markarfljót valley. A second gauge, downstream, recorded the flood moving at 2,500 cubic metres per second. Levees built for just such a moment held fast, channeling most of the flow away from farms. Thwarted, the churning waters rushed out through the valley and into the sea to the south. A second flood followed the next day, powerful enough to overtop the protective levees.

But that was just the water from Eyjafjallajökull. There was also, of course, the ash. As soon as earthquakes began rumbling beneath the ice-capped summit, scientists knew ash might be a problem. When hot magma meets cold ice, it generates steam that fragments the magma into zillions of tiny pieces. That's volcanic ash, most of which measures just a few millimetres across. Because the particles are so light, they can be hurled high into the atmosphere by the force of the eruption and then carried long distances by the winds.

Eyjafjallajökull began pouring out its infamous ash sometime in the early morning of 14 April. Icelandic officials first spotted it at 5.55 a.m. that day, from a survey plane sent to monitor the eruption. Those aboard saw a white plume rising through the cloud bank covering the mountain's summit. It was the first glimpse of the ash that would rewrite European rules on dealing with natural hazards.

Initially, the big challenge was to figure out how much ash was coming from Eyjafjallajökull and where it was going. Radar signals are used to obtain this information, but in April 2010 Iceland had just one weather radar system in operation, at the Keflavík complex, which sprawls on a peninsula west of Reykjavík. Once a United States air base, Keflavík is now the country's main commercial airport and thus a natural place to keep its main weather radar. But between Keflavík and Eyjafjallajökull lies a mountain range that blocked much of the radar's line of sight. For more than a quarter of the time that the volcano was erupting, the ash plume was too low for the Keflavík radar to see it. (Officials have since added a second radar in eastern Iceland, plus two mobile radars that can be driven where needed to track ash plumes.)

On that first day, the ash plume started out white and darkened throughout the afternoon as more and more material was spat out. By 6.30 p.m. it was nearly black and shot through with lightning. Westerly winds carried the ash eastward at first, towards northern Norway, where airspace closed that evening. The next day, the plume spread far wider and curled south, covering the airspace for other parts of Scandinavia as well as the United Kingdom.

This trajectory triggered a regional crisis. After the 1982 near-disaster, when a British Airways plane nearly crashed after flying through Indonesian ash, the International Civil Aviation Organization decided that no flights should go through any airspace where concentrations of volcanic ash are greater than zero. So as soon as Eyjafjallajökull began erupting, Icelandic scientists started sending their observations of the plume's location to the London Volcanic Ash Advisory Centre (VAAC). There are nine VAACs worldwide, each responsible for monitoring and forecasting volcanic ash plumes in their assigned region.

Icelandair names all its planes after the island's volcanoes, but surely didn't foresee the irony in this particular name.

Forecasters at the VAAC used computer simulations to predict where the Eyjafjallajökull plume might spread from day to day, depending on weather conditions and how much ash the volcano was ejecting. Day after day, those forecasts blocked out huge regions of airspace all across Europe. Aviation authorities followed the zero-tolerance rule for ash and shut down airports from Heathrow to Oslo. (Ironically, Keflavík itself remained open, since it lay west of the spreading plume.)

Some 100,000 flights were cancelled in the week after the eruption began, stranding ten million passengers. Only then did aviation officials begin reevaluating their definition of what was safe to fly in. Under huge pressure to get planes back in the air, governments started allowing flights to take off. In the end, the airspace closures cost Europe billions of euros because of lost passenger revenues and delays to air cargo. The consequences stretched as far as Kenya, where women were laid off by the floral industry after their flowers were left rotting at airports.

Eyjafjallajökull gave its final belch on 17 June 2010. All told, the volcano disgorged nearly 480 million tonnes of ash over more than seven million square kilometres of Europe and the North Atlantic. Less than 0.02 per cent of the erupted ash made it to mainland Europe – and yet even that modest amount caused utter chaos.

By most measures, Eyjafjallajökull wasn't an exceptional eruption. On the VEI scale of an eruption's power, it ranked a modest 3. Both Katla in 1918 and Grímsvötn in 2011 spat out more rock and ash than it did. What made Eyjafjallajökull so disruptive was a combination of three things: it erupted ash almost constantly for 39 days; its ash grains were relatively small, and so could be lofted a long way; and the winds carrying the ash just happened to be blowing from the north and northwest most of the time the mountain was erupting.

Some fifteen eruptions of Eyjafjallajökull's size go off in Iceland each century. But what might happen if a larger eruption were to occur tomorrow? What about, say, a Laki-style event? It would produce a lot less ash than Eyjafjallajökull did in 2010, but a lot more deadly gas. There's obviously no certainty here, but we can speculate on how such an eruption might unfold – and how it might change our world.

It all begins as small tremors start shaking south-central Iceland. At first the earthquakes are too tiny for anyone to feel, but the Icelandic Meteorological Office's seismic network picks them up, and a swarm of dots now show up on the IMO website. Let's say they appear just west of the Laki crater row, in the famous 'fire districts'.

A blogger picks up on the activity and writes an excited post about how Laki is reawakening. Almost immediately the IMO issues a statement correcting him on some factual errors, but the post is enough to draw the attention of the world's media. Newspapers and websites start running dire-sounding stories about an impending eruption, and the IMO website keeps crashing as people try looking up the activity for themselves. But gradually the earthquakes become less frequent, and then they disappear. It looks like the whole thing might have been a false alarm.

The volcano is only teasing. Nine years later, the small trembling of quakes reappears and suddenly accelerates its drumbeat. The quakes are clustering along a line that runs from the southwest to the northeast towards the huge Vatnajökull ice cap – plot the line on a map and it is nearly dead centre between the rift that opened in the year 939, in the Eldgjá eruption, and in 1783, as Laki.

Now slightly concerned, IMO scientists deploy more seismometers and global-positioning instruments to this sparsely settled area. Just in time, it seems: the newly installed GPS monitors discover that the western edge of the line of quakes is beginning to rise, and quickly. Magma must be on the move. The earthquakes are speeding up, and occurring at shallower depths.

A week and a half later, the ground rips open. Once again, the fire districts are aflame. Huge jets of lava shoot more than 1,000 metres in the air. Lava begins streaming from new volcanic cones and fissures, following the courses of rivers southward just as it did during the Laki eruption.

The well-prepared Icelanders spring into action. The town of Klaustur evacuates, along with farmers in the lava's path. Police set up checkpoints along the ring road, which lies directly in the course of the advancing lava. Tourists are warned not to travel eastward to the glaciers of Skaftáfell, one of the country's most popular national parks.

And then the clouds appear. Dark plumes billow above the fissure, laden with toxic gases. Once again, winds are just right

A farmer in Klaustur, masked against the dust clouds.

to start carrying the volcanic plume eastward and then toward the south. Scientists fly to the plume to measure its chemistry and calculate its likely course. Far overhead, satellites watch as the cloud stretches first toward Scandinavia, and then toward continental Europe and Great Britain.

When the volcanic haze drifts into Europe, people's eyes begin to itch and burn. Many have difficulty catching their breath. Environmental experts start measuring levels of sulphuric acid and fluorine in the haze, unsure if they can reassure people that the stuff is safe to breathe. Health authorities warn people not to spend too much time outdoors, especially old people and children.

Occasionally the winds shift, temporarily clearing the haze. But then they shift back, and the dry fog once again covers the countryside. The longer Iceland erupts, the less important the wind direction becomes. The haze has become so diffuse that it settles across Europe like a veil.

By this point airports in many parts of the continent are closed. Hospitals struggle to cope as emergency rooms are flooded with people struggling to breathe. The first deaths come among the old and infirm, and there is a huge run on N95 respirators, the kind that can filter out volcanic particles. No one had ever expected that huge swaths of Great Britain would need respirators, so health authorities have to triage and determine who needs them most urgently. A black market springs up on the Internet as people try to order breathing filters from other parts of the world – only to have deliveries delayed or stopped because ash has grounded flights.

Back in Iceland, agricultural experts warn about fluorine settling out from the haze. Farmers desperately seek to move their animals to unsullied pastures, but none can be found. Sheep die by the tens of thousands, and cattle suffering from fluorosis have to be put down. The country's livestock industry collapses.

The worst thing about the eruption is that it just goes on and on. For ten long months the fissure keeps spewing pollution. Eventually the ash encompasses so much of the northern hemisphere that all flights are cancelled for several months. Airline companies move as many planes as they can to Buenos Aires, Cape Town and Sydney, but many declare bankruptcy. Ships and trains struggle to move hospital supplies and food to the choking, hungry people of Europe.

Such a scenario may sound far-fetched, but it's not. 'I don't want to be a fearmonger,' says Anja Schmidt, 'but these eruptions happened and will happen again in Iceland. And we are more vulnerable now.' Modern society is so deeply dependent on transportation networks that it simply doesn't have enough resilience to deal with a Laki-like eruption. Such a cataclysm would test far more than whether jet engines can perform in volcanic ash. It would test to the limit, and probably break, public health and emergency response systems.

How likely is such a disaster? Not as unlikely as you might think. Although an Eyjafjallajökull-type eruption is far more common, a Laki-type eruption happens on average every 200 to 500 years. More than 200 years have passed since Laki blew up. Volcanoes don't erupt on predetermined cycles, of course, but they do show patterns of activity over time. Scientists can use those patterns to try to calculate how often eruptions of various sizes occur – all the way up to the planet-altering ones.

In the short term, the main thing to worry about is a fairly small eruption – of, say, a VEI 3 or 4 – going off relatively close to a city. Roughly 500 million people live close enough to an active volcano to be at risk. This includes cities such as Naples, close to Vesuvius; and Mexico City, not too far from the steaming Popocatépetl. (Popocatépetl's name is the Aztec word for 'smoking mountain'.) We've already examined the risks in Naples, so let's take a closer look at what Mexico's capital city might be facing.

Popocatépetl has unleashed its wrath three times in the last five millennia – Mesoamerican pottery and other artifacts entombed in the ancient mudflows bear witness to its terrible power. The last of these major eruptions took place around the year 800, and was powerful enough to create a plinian ash plume. At least one research team has argued that mudflows from this eruption buried almost everything around the great pyramid of Cholula, in the Puebla valley. (Some think the structure of the pyramid is a deliberate echo of the volcano's majestic peak.) Seven centuries later, when the conquistador Hernán Cortés reached Cholula, he marveled at the towering mountain from which 'both by day and night a great volume of smoke often comes forth and rises up into the clouds as straight as a staff, with such force that although a very violent wind continuously blows over the mountain range yet it cannot change the direction of the column.'

That's pretty much how Popo – as it's known locally – looks today. Its summit, at 5,246 metres above sea level, is frequently wreathed in clouds and ash, because after decades of slumber, Popo roared back to life in December 1994. Officials evacuated tens of thousands of people from the surrounding countryside. Since then Popo has continued to erupt intermittently, though not violently; in May 2013, it sent ash drifting over nearby towns.

If Popo were to go off with the ferocity of the 800 eruption, it would rank a VEI of 4. That would make it on the same scale as Laki, but its effects would be very different. The biggest threat would be not a choking haze but churning flows of mud, ash and other rocks, which would rush down the mountain's slopes and bury everything in their path.

That same sort of risk plagues another famous North American volcano: Mount Rainier, which towers 80 kilometres southeast of the urban sprawl of Seattle. In many ways Rainier is the archetype of the Cascades volcanoes, rising tranquilly above the lushly forested Pacific Northwest. But remember that Mount St. Helens is also in the Cascades, and it blew in 1980 with a VEI of 5, killing 57 people. It also sent the largest landslide in recorded history roaring down the river valley. Avalanching along at more than 200 kilometres an hour, it ripped bridges and houses from their foundations and reached its maximum size about 80 kilometres from the volcano itself. By the time it was all over, much of the river valley was buried under nearly 50 metres of debris.

There's nothing to say Rainier won't do the same. In fact, about 5,600 years ago it erupted much like Mount St. Helens. The blast sent mudflows coursing down Rainier's northeast slope, where they scoured everything in their path. The avalanches eventually covered about 550 square kilometres across the shores of Puget Sound, including the land that is now the port of Tacoma. Over time, other eruptions have piled fresh lava onto Rainier's summit, partially rebuilding it into a more symmetrical peak. Its last, relatively small, eruption happened about 1,000 years ago. Yet because it lies so close to so many towns, and because of its history of giant mudflows, Rainier remains the most dangerous volcano in the Cascades.

If Popo or Rainier goes off with a VEI of 4 or 5, people will be fairly well prepared. Both mountains are monitored pretty much constantly, and emergency managers are familiar with the kinds of evacuations they would have to conduct if it looked like mudslides were about to start rushing downhill. Undoubtedly not everything would go smoothly: there might be a false alarm or two, and some residents would ignore the warnings until it was too late. But Mexico City and Seattle would survive.

The picture changes when you start considering blasts in the range of VEI 6 and higher. These are on the order of Krakatau in 1883 and Tambora in 1815 – ones that cause regional disaster and have planet-wide effects. Over the past millennium, a blast up to the size of Krakatau has happened about twice per century. And there's a greater than 10 per cent chance that an even bigger, Tambora-style, event could go off in the next century.

For disasters on this scale, ground zero is likely to be somewhere in Asia. It's home to a quarter of the world's volcanoes, and more than two billion people. In 2012, Australian scientists took a broad look at 190 volcanoes across the Asia-Pacific region, aiming to calculate which ones were most likely to send ash drifting across the western part of the Ring of Fire.

Figuring out the eruption frequency of a given volcano might seem simple – take the number of times a volcano has gone off in the past and divide it by the time period you're studying. In practice, though, this approach is fraught with pitfalls. Notably, it's very hard to determine how often any volcano has erupted. Researchers tease this information out in various ways, such as by digging trenches to look for ash layers from past explosions, or by searching local archives

Evacuation routes for an eruption of Popocatépetl, Mexico, run through many small villages.

for accounts of major blasts. But it's more than possible for a major eruption to have occurred, comparatively recently, without leaving any written trace.

An unrecorded disaster of Tambora-like dimensions may in fact have happened in the year 1258. Volcanologists know that a VEI 7 eruption must have occurred at this time, because ice cores from both the Arctic and Antarctic show a spike in the amount of sulphur and ash in ice layers dating to that year or the year after. For eruption debris to have reached both poles, the volcano must have gone off somewhere in the tropics –

that is, at low enough latitudes for ash and sulphur aerosols to be mixed through the atmosphere and across the equator.

An explosion of this size would have altered climate worldwide, and there seems to be evidence that just such a change did indeed happen. In 2012, a Museum of London archaeology team reported that 10,000-plus bodies found in a mass grave by East London's Spitalfields Market might date to the year 1258. The scientists can't pinpoint the exact date of the cemetery, but they have found documentary records of heavy rains, crop failure and famine around 1258, which does add support to the theory that the Londoners died in the aftermath of a colossal eruption.

But what volcano could have been responsible? Some of the top contenders have included El Chichón, in Mexico, and Ecuador's Quilotoa, but the chemistry of their magma doesn't match the ash and sulphur found in the ice cores from 1258. What now looks most likely is that the eruption happened in Indonesia, the planet's most active volcanic region – specifically at the Rinjani volcano, in the Lesser Sunda Islands. Rinjani has an 8.5-kilometre-wide caldera that is known to have been formed some time in the thirteenth century.

It's hard to believe that such a massive blast could have occurred, not all that long ago, and yet have left no mark in the historical record – and yet this appears to be what has happened. The mystery of the 1258 mega-explosion underscores our ignorance, and this is what is worrying. Who knows if another Indonesian volcano, dormant for centuries and unremarked, is ready to suddenly and calamitously explode?

Any tabulation of the world's most dangerous volcanoes wouldn't be complete without a visit to the supervolcanoes – those of VEI 8 or greater, which spew more than a thousand cubic kilometres of material into the sky. The biggest of these eruptions in the past few hundred thousand years was Toba, 74,000 years ago. The most recent was in New Zealand about

26,000 years ago, in the Taupo volcanic zone. That involved ten separate eruptions over several years, which ultimately buried most of New Zealand's North Island in its volcanic vomit.

If such a supereruption were to occur in Trafalgar Square, it would bury Greater London 700 metres deep in ash. But it doesn't really matter where exactly the volcano goes off: any supereruption would devastate the planet. Within weeks, the veil of erupted particles would screen out the sun, freezing and killing almost all vegetation. The searing blast would destroy most of the atmosphere's protective ozone layer, exposing all surviving plants and animals to a flood of deadly ultraviolet radiation. In the longer term, the eruption's sulphate aerosols would plunge the Earth into an unrelenting volcanic winter, lowering temperatures by as much as 10 degrees Celsius for a decade. There is no precedent for this kind of global natural disaster.

A supereruption occurs perhaps once every 50,000 years. Such an eruption would be what economists call a 'black swan' event – one that has an extreme impact on human lives but is almost impossible to predict. But improbability is of course not the same as impossibility. Think of the flooding of the Fukushima nuclear plant in Japan in March 2011, for example – a textbook 'black swan' disaster.

So what can we do? For starters, we can step up monitoring at plausible candidates for future supereruptions. Volcano monitoring got its start in 1841 at Vesuvius, when the King of Naples established the Osservatorio Vesuviano halfway up the volcano's western slope. This was the only such scientific outpost for more than seven decades, until Thomas Jaggar established his at Kilauea in Hawaii. Today there are about 100 volcano observatories and research institutes worldwide, from the Philippines to Kamchatka to Alaska.

The backbone of volcano monitoring is seismology. Magma shifting in deep reservoirs causes the ground to move, a shift that can be detected by seismometers. Volcanic earthquakes are very different from ordinary earthquakes, though; around

The saucer-shaped object on top of the tripod is a GPS antenna that measures if the ground (here, at Yellowstone) is deforming and possibly indicating a future eruption.

volcanoes they tend to come in swarms of tiny quakes, with magnitudes typically less than 2. You would never feel these quakes even if you were standing right on the volcano as it went off. The quakes also come in two classes: higher-frequency ones generated when magma moves and fractures rock, and lower-frequency ones that are usually caused by gas bubbles forming and popping in the magma. A low-frequency 'tremor' also often shows up before an eruption, probably generated by magma flowing within the ground.

All these signals are sent to scientists at the observatory, who then have to make a decision about what they actually mean.

Sometimes, but not always, an earthquake swarm signals an impending eruption. If the high-frequency quakes suddenly disappear, that could be another sign that an eruption is imminent. And the energy seen in volcanic tremors, before an eruption starts, can indicate how much material is going to be released.

There are other ways to feel a volcano's heartbeat. Global-positioning and other instruments measure changes in the shape of the ground. Gas-monitoring devices stuck into volcanic vents can record changes in the chemistry of volcanic gases. And satellites keep an eye on all the world's volcanoes from the sky, sometimes providing the only evidence of very remote eruptions, thanks to an ash plume visible in satellite images.

Many of the world's potential supervolcanoes, such as Yellowstone, are under this kind of close watch. None of them show any hint of unleashing a VEI 8 blast any time soon. Still, there are probably plenty of unknown candidates – especially, as noted, in Indonesia. If one of these underappreciated volcanoes decides to blow, scientists may not have much advance warning at all.

As if that weren't enough to worry about, the chances of an eruption in colder parts of the world, such as Iceland, also rise every day, thanks to climate change. As Iceland's ice has melted, its volcanoes have become more active, because the overlying weight of ice has been reduced. This has been going on since the last ice age ended around 12,000 years ago. Great ice sheets that had scraped southward into Europe and North America then retreated, and landscapes that had been buried by tonnes of ice began to rise. Parts of Scandinavia are still uplifting because of this phenomenon, known as postglacial rebound.

Superimposed on the natural rebound is the effect of human activity. Every day, the burning of fossil fuels adds more heat-trapping greenhouse gases to the atmosphere. As a result, ice is melting all the faster. The great ice sheets atop Greenland and Antarctica are shedding fresh water into the oceans at an accelerating rate; their meltwater accounts for about one-third of the sea level rise seen in recent years. (Much of the rest of the rise comes from the fact that water expands as it gets warmer.)

Iceland, too, is melting. Since 1890, the island has lost 435 cubic kilometres of ice, primarily from the Vatnajökull ice cap. The loss of this ice is changing the geological stresses in the crust beneath, in ways that have altered how much magma gets made – it's simple cork-and-champagne physics. Geologists estimate that the amount of magma produced beneath Iceland has increased thirty-fold since the end of the last ice age. In the past century alone, magma production rates may have gone up by as much as fifteen per cent. In other words, there is now more magma available to feed volcanoes such as Laki.

A Laki-style eruption, then, may be a far from improbable occurrence. Would it be a disaster? Compared to the global catastrophe of a supereruption, a VEI 4 eruption on the scale of Laki might seem almost tame. But a moderate-sized blast is far more likely to occur in the next few centuries than a Yellowstone-size behemoth, and the consequences of such an explosion – as we have seen – could be grave.

EPILOGUE

Return to Heimaey

THERE IS JUST ONE CHURCH on the tiny island of Heimaey, a simple white building with a charcoal-gray steeple. To reach it you turn your back to the water and your face toward a newborn volcano, and you walk uphill. On the evening of 22 March 1973, nearly everyone who was still on Heimaey made this walk.

The battle to save the island's port was in top gear, and it was going badly. Spraying seawater on the advancing lava worked to cool it, but not as much as engineers had hoped. Shifts of workers pumped around the clock, and still the lava came. The islanders decided to appeal to a higher power. Thus came about the second Fire Mass.

More so than the rest of Iceland, Heimaey is a fairly religious community, perhaps because of its history of loss. In 1627, pirate ships attacked the island, kidnapping some three-quarters of the residents and murdering most of the rest. In the latter part of the nineteenth century, Heimaey again lost a fair chunk of its population when about 200 islanders converted to Mormonism and followed missionaries to Utah. Those who remained drew together closer than ever. Heimaey, after all, means 'home island' in Icelandic.

So when it seemed that the 1973 eruption was about to bury their town in lava, the islanders turned to the church. Surely, they thought, it couldn't hurt to come together and renew the fight with a little spiritual backing. Thorsteinn Lúther Jónsson, one of the town's two priests, agreed to conduct the service.

That evening, nearly everybody on the rescue teams crowded into the church. Grim-faced men, wearing traditional lopapeysa sweaters, came to hear Thorsteinn. They lit long tapers to illuminate the dim church, even as the lava fountains sprayed brilliant orange fire outside.

It didn't work. Despite the prayers of hundreds of people, a new lava flow broke out that night and headed directly for the town. Within eight hours 64 houses were burned or buried under lava, and the following day 30 more homes were destroyed. That was when islanders started to joke with Pastor Thorsteinn about how 'his mass' didn't work as well as Jón Steingrímsson's mass in Klaustur. He is said not to have found that funny at all.

Perhaps Heimaey's fire mass wasn't heartfelt enough. Or perhaps Pastor Jón just got lucky two centuries ago. There is no obvious way to cope when your town is on the brink of destruction, and it's probably worth trying every weapon in the arsenal – from seawater spraying to divine intervention – to save it. The planet is not going to rein in its volcanic fire on our behalf.

Endnotes

INTRODUCTION

The classic narrative of Heimaey's 1973 eruption lies in John McPhee's *The Control of Nature* (in a piece originally written for *The New Yorker*). A vivid description from the islanders' perspective appears in the 2003 issue of the now-defunct *Icelandic Geographic*, by Hjálmar R. Bárdarson and Margrét Jónasdóttir. For accounts of the engineering that stopped the lava, we relied on *U.S. Geological Survey Open-File Report* 97–724, in which Richard S. Williams, Jr., collected English translations of some of the key Icelandic descriptions of the lava-cooling operations. More technical details of the eruption appear in a USGS booklet, *Man Against Volcano*, originally published in 1976 and written by the same author along with James G. Moore. Additional context can be found in 'The Eldfell Eruption, Heimaey, Iceland: A 25-Year Retrospective,' by Alan V. Morgan in the March 2000 issue of *Geoscience Canada*.

CHAPTER 1

The writings of Jón Steingrímsson have been translated into English in recent decades. His famous chronicle (of which there were actually three versions) appears as *Fires of the Earth: The Laki Eruption 1783–1784*, translated by Keneva Kunz and published in 1998 by the Nordic Volcanological Institute and the University of Iceland Press. Michael Fell's 2002 translation of Jón's autobiography, *A Very Present Help in Trouble: The Autobiography of the Fire-Priest*, from Peter Lang Publishing, was of great help in understanding Jón's childhood and world view. For statistics on Laki's lava flows and gas emissions, we referred to the seminal work of Thorvaldur Thordarson and Stephen Self, which will be referenced many times to come: notably 'Atmospheric and environmental effects of the 1783–1784 Laki eruption: a review and reassessment,' which appeared in the *Journal of Geophysical Research* in 2003.

CHAPTER 2

Haraldur Sigurdsson's *Melting the Earth: The History of Ideas on Volcanic Eruptions* (Oxford University Press, 1999) is notable not only for its scholarship but also the extensive collection of volcano images from the author's personal collection. For more on the concept of plate tectonics, see *Plate Tectonics: An Insider's History of the Modern Theory of the Earth*, edited by Naomi Oreskes (Westview Press, 2001). Alfred Wegener's seminal 1915 publication was *The Origin of Continents and Oceans*, available in several English-language translations.

Details on Iceland's geologic past (including the Katla magic trousers story, plus many volcano statistics) come from the thorough field guide that is part of the *Classic Geology in Europe* series: *Iceland* by Thor Thordarson and Armann Hoskuldsson (Terra Publishing, 2002). *Hekla on Fire* by Sigurdur Thorarinsson (Hanns Reich Verlag, 1956) describes the history and geology of Iceland's most famous volcano; see also *The History of Iceland* by Gunnar Karlsson (University of Minnesota Press, 2000). Katla 1918 flood calculations come from John A. Stevenson of the University of Edinburgh on his invaluable blog, *volcan01010* (all-geo.org/volcan01010).

CHAPTER 3

For an overview of Toba research, see the papers collected as volume 258 of *Quaternary International* (2012). For Santorini, see Floyd W. McCoy and Grant Heiken, 'The Late Bronze Age eruption of Thera (Santorini), Greece: regional and local effects,' *Geological Society of America Special Paper 345* (2000), as well as Andrew V. Newman et al., 'Recent geodetic unrest at Santorini Caldera, Greece,' *Geophysical Research Letters*, vol. 39, L06309 (2012).

Vesuvius is well represented in many scholarly works, including a 1982 overview from Haraldur Sigurdsson et al., 'The eruption of Vesuvius in A.D. 79: reconstruction from historical and volcanological evidence,' *American Journal of Archaeology*, vol. 86, pp. 39–51, and Gillian Darley's *Vesuvius* (Profile Books, 2011). Also see Richard B. Stothers and Michael R. Rampino, 'Volcanic eruptions in the Mediterranean before A.D. 630 from written and archaeological sources,' *Journal of Geophysical Research*, vol. 88, pp. 6357–6371 (1983). For more recent research see, for instance, Lucia Pappalardo and Giuseppe Mastrolorenzo, 'Short residence times for alkaline Vesuvius magmas in a multi-depth supply system,' *Earth and Planetary Science Letters*, vol. 296, pp. 133–143 (2010).

Papers on the global impact of Tambora are collected in a workshop volume edited by C.R. Harington, *The Year Without a Summer?: World Climate in 1816* (Canadian Museum of Nature, 1992). Sir Thomas Raffles's *History of Java* is readily available online. For insight into the Indonesian records and beliefs, see Bernice de Jong Beers, 'Mount Tambora in 1815: a volcanic eruption in Indonesia and its aftermath,' Indonesia, pp. 37–60 (1995). The Alsace farmer's quote comes from Tom Bodenmann et al., 'Perceiving, explaining and observing climatic changes: an historical case study of the "year without a summer" 1816,' *Meteorologische Zeitschrift*, vol. 20, pp. 577–587 (2011). The Irish doctor's quote is cited in Clive Oppenheimer, 'Climate, environmental and human consequences of the largest known historical eruption: Tambora volcano (Indonesia) 1815,' *Progress in Physical Geography*, vol. 27, pp. 230–259 (2003). W.J. Humphreys's observations on volcanic dust appear in vol. 6 of the *Bulletin of the Mount Weather Observatory* (1913).

Everything you ever wanted to know about Krakatau is in the exhaustingly complete *Krakatau 1883: The volcanic eruption and its effects* by Tom Simkin and Richard S. Fiske (Smithsonian Institution Press, 1983). Also see the original Royal Society report *The eruption of Krakatau and subsequent phenomena*, edited by G.J. Symons (London, 1888).

CHAPTER 4

See the notes for Chapter 1 for many of the sources used here. Once again, Thor Thordarson has done much of the seminal work; for instance, there are actually three *Eldrits* written by Jón Steingrímsson, and Thordarson compares them meticulously in the hard-to-find journal *Jökull*, number 53 (2003).

Magnús Stephensen's account appears in English translation in the second edition of William Jackson Hooker's *A Tour in Iceland in the Summer of 1809* (London, 1813).

CHAPTER 5

Accounts of Laki's European effects are collected in Thorvaldur Thordarson and Stephen Self, 'Atmospheric and environmental effects of the 1783–1784 Laki eruption: a review and reassessment,' *Journal of Geophysical Research*, vol. 108, D1 (2003). Gilbert White's *The Natural History and Antiquities of Selborne* can be found in various printings; we relied on The Folio Society version (1994). His journals can be found at *naturalhistoryofselborne.com*.

Richard Stothers wrote an overview of 'the great dry fog of 1783' in *Climatic Change*, vol. 32, pp. 79–89 (1996). William Herschel's journals are collected digitally by the Royal Astronomical Society. Professor Van Swinden's account, translated by Susan Lintleman and commented on by Thordarson and Self, appears in *Jökull*, number 50, pp. 65–80 (2001).

Public perceptions of the dry fog are collected by John Grattan and Mark Brayshay in 'An amazing and portentous summer: environmental and social responses in Britain to the 1783 eruption of an Iceland volcano,' *The Geographical Journal*, vol. 161, pp. 125–134 (1995). A wider-ranging paper of Grattan and Brayshay, exploring the effects in all of Europe, appeared in *Volcanoes in the Quaternary*, edited by C.R. Firth and W.J. McGuire, Geological Society Special Publication 161, pp. 173–187 (1999). Also see Sven Laufeld, 'The Lakagígar 1783–84 eruption and its environmental impact in the Nordic countries,' *GFF*, vol. 116, p. 211–214 (1994). One Leeds Intelligencer quote is sourced from Alwyn Scarth, *Vulcan's Fury: Man Against the Volcano* (Yale University Press, 1999).

For more on the year of wonders, see 'Meteors and perceptions of environmental change in the annus mirabilis AD1783–4' by Richard J. Payne, *North West Geography*, vol. 11, pp. 19–28 (2011). For more on mortality, see C.S. Witham and C. Oppenheimer, 'Mortality in England during the 1783–4 Laki Craters eruption,' *Bulletin of Volcanology*, vol. 67, pp. 15–26 (2005), as well as John Grattan et al., 'Volcanic air pollution and mortality in France 1783–1784,' *Comptes Rendus Geoscience*, vol. 337, pp. 641–651 (2005).

The Marie Antoinette story, and details on the European floods, come from Rudolf Brázdil et al, 'European floods during the winter 1783/1784: scenarios of an extreme event during the "Little Ice Age"', *Theoretical & Applied Climatology*, vol. 100, pp. 163–189 (2010). Also see Emmanuel Garnier, 'La ville face aux caprices du fleuve,' *Histoire Urbaine*, number 18, pp. 41–60 (2007).

A wealth of US weather information can be found in David M. Ludlum's *Early American Winters 1604–1820* (American Meteorological Society, 1966).

CHAPTER 6

Franklin's ruminations on the dry fog can be found in his *Meteorological Conjectures*. Also see Richard J. Payne, 'The Meteorological Imaginations and Conjectures of Benjamin Franklin,' *North West Geography*, vol. 10, pp. 1–7 (2010).

For more on the debate over who first linked the dry fog to an Icelandic eruption, see Sigurdur Thorarinsson, 'Greetings from Iceland: ash-falls and volcanic aerosols in Scandinavia,' *Geografiska Annaler*, vol. 63, pp. 109–118 (1981).

The Labrador haze sighting appears in Gaston R. Demarée et al, 'Further documentary evidence of northern hemispheric coverage of the great dry fog of 1783,' *Climatic Change*, vol. 39, pp. 727–730 (1998). The possible Inuit link is described in G.C. Jacoby et al, 'Laki eruption of 1783, tree rings, and disaster for northwest Alaska Inuit,' *Quaternary Science Reviews*, vol. 18, pp. 1365–1371 (1999). Reports of southern hemisphere observations are in Ricardo M. Trigo et al, 'Witnessing the impact of the 1783–1784 Laki eruption in the Southern Hemisphere,' *Climatic Change*, vol. 99, pp. 535–546 (2010).

Thordarson's calculations of fluorine content in the Laki magma are in 'Sulfur, chlorine, and fluorine degassing and atmospheric loading by the 1783–1784 Laki (Skáftar Fires) eruption in Iceland,' *Bulletin of Volcanology*, vol. 58, pp. 205–225 (1996).

Anyone interested in the question of whether Laki's aerosols really reached the stratosphere can check out a pair of competing papers, left out here for reasons of space and clarity. One argues against a stratospheric impact: Alyson Lanciki et al, 'Sulfur isotope evidence of little or no stratospheric impact by the 1783 Laki volcanic eruption,' *Geophysical Research Letters*, vol. 39, L01806 (2012). Countering that is Anja Schmidt et al, 'Climatic impact of the long-lasting 1783 Laki eruption: inapplicability of mass-independent sulfur isotopic composition measurements,' *Journal of Geophysical Research*, vol. 117, D23116 (2012). We, naturally, side with the latter.

The modelling work of Robock and others appears in two main papers: Luke Oman et al, 'Modeling the distribution of the volcanic aerosol cloud from the 1783–1784 Laki eruption,' *Journal of Geophysical Research*, vol. 111, D12209 (2006), and Oman et al, 'High-latitude eruptions case shadow over the African monsoon and the flow of the Nile,' *Geophysical Research Letters*, vol. 33, L18711 (2006). For a possible modern analog to the winter that followed Laki, see Rosanne D'Arrigo et al, 'The anomalous winter of 1783–1784: was the Laki eruption or an analog of the 2009–2010 winter to blame?', *Geophysical Research Letters*, vol. 38, L05706 (2011). For more on historical European climate, see Brian Fagan, *The Little Ice Age* (Basic Books, 2000). For a possible link to volcanoes, see Gifford H. Miller et al, 'Abrupt onset of the Little Ice Age triggered by volcanism and sustained by sea-ice/ocean feedbacks,' *Geophysical Research Letters*, vol. 39, L02708 (2012).

CHAPTER 7

Most of this chapter is based on our June 2012 trip to Klaustur and its environs. For an interesting citizen-science angle on the 2011 Grímsvötn eruption, in which UK residents collected ash samples on sticky tape to generate a map of ash dispersal, see John A. Stevenson et al, 'UK monitoring and deposition of tephra from the May 2011 eruption of Grímsvötn, Iceland,' *Journal of Applied Volcanology*, vol. 2, number 3 (2013).

CHAPTER 8

Survivor stories from the Lake Nyos disaster are collected in F. Le Guern et al, 'Witness accounts of the catastrophic event of August 1986 at Lake Nyos (Cameroon),' *Journal of Volcanology and Geothermal Research*, vol. 51, pp. 171–184 (1992). For myths of the exploding lakes, see Eugenie Shanklin, 'Exploring lakes and maleficent water in Grassfields legends and myth,' *Journal of Volcanology and Geothermal Research*, vol. 39, pp. 233–246 (1989). For an overview of degassing efforts, see Michel Halbwachs et al, 'Degassing the "killer lakes" Nyos and Monoun, Cameroon,' *EOS*, vol. 85, pp. 281&285 (2004).

Estimates on total number of volcanic fatalities are fraught with error, but for one attempt see Tom Simkin et al, 'Volcano fatalities: lessons from the historical record,' *Science*, vol. 291, p. 255 (2001). We also relied on the excellent reference *Volcanic Hazards: A Sourcebook on the Effects of Eruptions* by R.J. Blong (Academic Press Australia, 1984). Estimated fatality numbers are recorded in Lee Siebert, Tom Simkin and Paul Kimberly's *Volcanoes of the World* (University of California Press, third edition, 2010).

A good overview of Iceland's volcanic risks can be found in Magnús T. Gudmundsson et al, 'Volcanic hazards in Iceland,' *Jökull*, vol. 58, pp. 251–268 (2008). More on the 1996 Grímsvötn eruption, often known by the name of the Gjálp fissure that erupted, is in Gudmundsson et al, 'Ice-volcano interaction of the 1996 Gjálp subglacial eruption, Vatnajökull, Iceland,' *Nature*, vol. 389, pp. 954–957 (1997).

Hawaii's vog dangers appear, among other places, in Bernadette M. Longo et al, 'Acute health effects associated with exposure to volcanic air pollution (vog) from increased activity at Kilauea volcano in 2008,' *Journal of Toxicology and Environmental Health*, vol. 73, pp. 1370–1381 (2010). Longo reported the asthma numbers at the Cities on Volcanoes conference in Colima, Mexico, in November 2012.

Fluorine in the Laki magma chamber is described in Maryjo Brounce et al, 'Insights into crustal assimilation by Icelandic basalts from boron isotopes in melt inclusions from the 1783–1784 Lakagígar eruption,' *Geochimica et Cosmochimica Acta*, vol. 94, pp. 164–180 (2012). Fluorine estimates for Eyjafjallajökull are in E. Bagnato et al, 'Scavenging of sulphur, halogens and trace metals by volcanic ash: the 2010 Eyjafjallajökull eruption,' *Geochimica et Cosmochimica Acta*, vol. 103, pp. 138–160 (2013).

Egil's Saga makes for great reading on its own, but for fluorine details see Philip Weinstein, 'Palaeopathology by proxy: the case of Egil's bones,' *Journal of Archaeological Science*, vol. 32, pp. 1077–1082 (2005). The 2004 exhumation project is described by Hildur Gestsdóttir and colleagues in *Fluorine poisoning in victims of the 1783–84 eruption of the Laki fissure, Iceland* (Fornleifastofnun Íslands, Reykjavík, 2006).

Witham and Oppenheimer (2005) addressed the two mortality peaks in England after the Laki eruption. The Donora smog is described in James G. Townsend, 'Investigation of the smog in Donora, Pa., and vicinity,' *American Journal of Public Health*, vol. 40, pp. 183–189 (1950).

Mayor Ken Livingstone's office published a descriptive booklet in 2002 on '50 years on: The struggle for air quality in London since the great smog of December 1952,' with some historical detail. Other figures come from Michelle L. Bell et al, 'A retrospective assessment of mortality from the London smog episode of 1952: the role of influenza and pollution,' *Environmental Health Perspectives*, vol. 112, pp. 6–8 (2004), and Andrew Hunt et al, 'Toxicologic and epidemiologic clues from the characterization of the 1952 London smog fine particular matter in archival autopsy lung tissues,' *Environmental Health Perspectives*, vol. 111, pp. 1209–1214 (2003).

Laki's health effects are addressed in many publications by John Grattan, but for historical reports here we relied on Michael Durand and Grattan, 'Extensive respiratory health effects of volcanogenic dry fog in 1783 inferred from European documentary sources,' *Environmental Geochemistry and Health*, vol. 21, pp. 371–376 (1999). For an overview, also see *Living Under the Shadow: The Cultural Impacts of Volcanic Eruptions*, edited by John Grattan and Robin Torrence (Left Coast Press, 2007).

Anja Schmidt's work appeared in Schmidt et al, 'Excess mortality in Europe following a future Laki-style Icelandic eruption,' *Proceedings of the National Academy of Sciences*, vol. 108, pp. 15710–15715 (2011).

CHAPTER 9

There's no shortage of publications on the many aspects of the 2010 Eyjafjallajökull eruption, but the Icelandic Meteorological Office, the University of Iceland Institute of Earth Sciences, and the National Commissioner of the Icelandic Police compiled an authoritative report in June 2012 for the International Civil Aviation Organization. It is available via the IMO's English-language web site, http://en.vedur.is. Also see Magnús T. Gudmundsson et al, 'Ash generation and distribution from the April–May 2010 eruption of Eyjafjallajökull, Iceland,' *Scientific Reports*, vol. 2, 572 (2012), as well as F. Sigmundsson et al, 'Intrusion triggering of the 2010 Eyjafjallajökull explosive eruption," *Nature*, vol. 468, pp. 426–430 (2010).

For rates of Icelandic eruption over time, see T. Thordarson and G. Larsen, 'Volcanism in Iceland in historical time: volcano types, eruption styles and eruptive history,' *Journal of Geodynamics*, vol. 43, pp. 118–152 (2007).

For Popocatépetl, see Claus Siebe et al, 'Repeated volcanic disasters in Prehispanic time at Popocatépetl, central Mexico: past key to the future?', *Geology*, vol. 24, pp. 399–402 (1996). Archaeological digs at the site are described in Patricia Plunket and Gabriela Uruñuela, 'Appeasing the volcano gods,' *Archaeology*, vol. 51 (1998).

For Mount Rainier, see James W. Vallance et al, 'Debris-flow hazards caused by hydrologic events at Mount Rainier, Washington,' *USGS Open-File Report* 03–368 (2003). For Mount St. Helens, see Vallance et al, 'Mount St. Helens: A 30-year legacy of volcanism,' *EOS*, vol. 91, pp. 169–170 (2010).

Steve Self warned about VEI 6–7 events at the International Association of Volcanology and Chemistry of the Earth (IAVCEI) conference in July 2013 in Kagoshima, Japan.

For Asian volcanic risk, see Susanna Jenkins et al, 'Regional ash fall hazard I: a probabilistic assessment methodology,' *Bulletin of Volcanology*, vol. 74, pp. 1699–1712 (2012). For the mystery eruption in 1258, see Richard B. Stothers, 'Climatic and demographic consequences of the massive volcanic eruption of 1258,' *Climatic Change*, vol. 45, pp. 361–374 (2000). The Spitalfields excavation, and the possible link to a volcanic eruption, are detailed in the Museum of London Archaeology monograph 'A bioarchaeological study of medieval burials on the site of St. Mary Spital: excavations at Spitalfields Market, London E1, 1991–2007" (2012). For the Rinjani identification, see F. Lavigne et al, 'Source of the great A.D. 1257 mystery eruption unveiled, Samalas volcano, Rinjani volcanic complex, Indonesia,' *Proceedings of the National Academy of Sciences*, doi:10.1073/pnas.1307520110 (2013).

Supereruptions lend themselves to dramatic sounding papers, such as Michael R. Rampino, 'Supereruptions as a threat to civilizations on earth-like planets,' *Icarus*, vol. 156, pp. 562–569 (2002). Also see S. Self, 'The effects and consequences of very large explosive volcanic eruptions,' *Philosophical Transactions of the Royal Society A*, vol. 364, pp. 2073–2097 (2006), and S. Sparks et al, 'Super-eruptions: global effects and future threats,' *Report of a Geological Society of London Working Group* (2005).

Monitoring details are sourced from John P. Lockwood and Richard W. Hazlett, *Volcanoes: Global Perspectives* (Wiley-Blackwell, 2010). On the topic of climate change and magma generation, see Carolina Pagli and Freysteinn Sigmundsson, 'Will present day glacier retreat increase volcanic activity? Stress induced by recent retreat and its effect on magmatism at the Vatnajökull ice cap, Iceland,' *Geophysical Research Letters*, vol. 35, L09304 (2008), as well as Hugh Tuffen, 'How will melting of ice affect volcanic hazards in the twenty-first century?' *Philosophical Transactions of the Royal Society A*, vol. 368, pp. 2535–2558 (2010). For more on this general topic, see Bill McGuire, *Waking the Giant: How a Changing Climate Triggers Earthquakes, Tsunamis and Volcanoes* (Oxford University Press, 2012).

EPILOGUE

Details of the Heimaey fire mass came via Helga Hallsbergsdóttir, of Heimaey, and the photographs of Sigurgeir Jónsson.

Photo Credits

All maps created by Theresa Dubé © Author

TITLEPAGE

Hekla engraving © Depositphotos

INTRODUCTION

Man vs lava © Sigurgeir Jónasson

Heimaey's harbour © Author photo

CHAPTER 1

Mauna Loa eruption © J.D. Griggs, US Geological Survey

Church in Skógar © Author photo

Laki crater © Author photo

Church ornaments © Author photo

Eyjafjallajökull lightning © Snaevarr Gudmundsson/Getty Images

CHAPTER 2

Wegener/Villumsen © Archive for German Polar Research/Alfred Wegener Institute

Mid-ocean ridges © Wikimedia Commons

Tectonic plates © US Geological Survey

Subducting ocean plates © US Geological Survey

Mid-Atlantic Rift © US Geological Survey

Ortelius's map of Iceland © helmink.com

Katla's 1918 eruption © University of Iceland Institute of Earth Sciences

May 11 2010 ash plume from Eyjafjallajökull © NASA

Grímsvötn August 2011 © Henrik Thorburn/Wikimedia Commons

PHOTO CREDITS

CHAPTER 3

Yellowstone geyser © Author photo
Toba space shot © Google Earth
Santorini overview © Google Earth
Pompeii fresco © Mark Ellingham
Tambora from space © NASA Earth Observatory
Krakatau eruption lithograph © Royal Society

CHAPTER 4

Prestbakki altar © Author photo
Laki lava flow map © Author sketch, after Fig. 7 in Thordarson & Self 1993
Sheep farmers after Grímsvötn eruption © Getty Images
Laki lava flows with moss © Author photo

CHAPTER 5

Herschel 20-foot telescope © The Scientific Papers of Sir William Herschel (1912)
August 1783 meteor © The Hunterian, University of Glasgow 2013
Erban Prague engraving © Prague City Museum

CHAPTER 6

Benjamin Franklin engraving © Library of Congress
NICL deep freezer © US Geological Survey
Nile river drops, from Oman 2006 © Courtesy Alan Robock; graphic from Oman et al 2006, based on data from Kondrashov et al 2005
Volcanic atmospheric cooling © Berkeley Earth

CHAPTER 7

Crater row northeast © Author photo
Klaustur's memorial chapel © Author photo
Jón Helgason © Author photo
Leading a pony through ash from Grímsvötn © AFP/Getty Images
Page from Book of Fire © Author photo

CHAPTER 8

Lake Nyos degassing © Bill Evans, US Geological Survey
Unzen pyroclastic flows © T. Kobayashi/University of Kagoshima; US Geological Survey

Pele, the Hawaiian fire goddess © Author photo
Egil Skallagrímsson © Wikimedia Commons
London smog © Popperfoto/Getty Images

CHAPTER 9

Icelandair plane 'Eyjafjallajökull' © AFP/Getty Images
Klaustur farmer with mask © Ingolfur Juliusson/Reuters/Corbis
Evacuation route for Popocatépetl, Mexico © Hector Aiza Ramirez/
Demotix/Corbis
GPS antenna at Porkchop, Yellowstone © USGS

Acknowledgements

This book would not have been possible without the generous help of many. To begin with we must acknowledge Lindy Elkins-Tanton, who over dinner at a Boulder restaurant some years ago made the fatal comment of suggesting that a book about Laki would be a good idea.

We are also indebted to the scholarship of true Laki experts, foremost among them Thorvaldur Thordarson of the University of Iceland. Thordarson's work on the Laki eruption – its details and consequences, both scientific and cultural – is unmatched, and you will find many of his papers referenced in the endnotes. He also shared many of his experiences in an in-person interview. Thordarson's former advisor, Stephen Self of the US Nuclear Regulatory Commission, also helped with references and general advice on world-changing eruptions. We relied heavily on the published research of John Grattan, of Aberystwyth University, for the environmental and health effects of the eruption.

Among the younger generation of Icelandic volcano experts, we are particularly indebted to Anja Schmidt of the University of Leeds and John Stevenson of the University of Edinburgh. Both endured lengthy sets of pestering questions, and both graciously provided technical comments on portions of the manuscript. Any errors that remain are, of course, ours.

In Iceland, England and the United States, we benefitted from the advice of many volcanologists and historians. Among them, and in no particular order, we thank Haraldur Sigurdsson, Freysteinn Sigmundsson, Benedikt Ofeigsson, Bergrún Óladóttir, Sveinbjörn Rafnsson, Helga Hallbergsdóttir, Jón Torfason, Astrid Ogilvie, Alan Robock, Rosanne D'Arrigo, Peter Baxter, Emmanuel Garnier, Richard Payne, Rudolf Brázdil, Michael Fell, Ian Skilling, Emily Constantine Mercurio, Hugh Tuffen, Ben Edwards, John Maclennan, Sue Loughlin, Stephen Sparks and Ken Carslaw.

Dave McGarvie was immensely helpful for Icelandic logistics and background information.

In Klaustur itself, we thank Jón Helgason, the local keeper of the Laki flame, and Sveinn Jensson of the Icelandair Hotel Klaustur for facilitating much of our visit. Hólasport provided transportation to the Laki crater row. The National and University Library of Iceland allowed us access to one of the original Eldrit manuscripts. Geoff Hargreaves brought out the Laki sample for us at the National Ice Core Laboratory in Denver.

A journalism fellowship from the European Geosciences Union helped with travel costs. The University of Washington's Friday Harbor Laboratories provided us a tranquil writing space in their Helen Riaboff Whiteley Center in the San Juan Islands.

William West, Carol Witze, Brendan Borrell and Mason Inman provided useful feedback on the writing. Jim Williams, Chris Witze, Torben Brun and Elle Jauffret helped with translation issues. An unnamed librarian at the USGS library in Reston, Virginia, has our eternal thanks for scanning the Helland map of the crater row, and Kathleen Cassaday and her colleagues at the Boulder Public Library efficiently fulfilled many requests for obscure Icelandic documents. Colleagues at *Science News* and *Nature* magazines put up with Alex's obsessions with all things volcanic, and Twitter colleagues including but not limited to Erik Klemetti provided an excellent sounding board for virtual volcanological discussions.

We thank Anna Carmichael at the Abner Stein literary agency, and especially Jeff's agent Regula Noetzli who launched the project into reality. Jonathan Buckley provided insightful line editing, and Mark Ellingham of Profile Books championed the idea from the start. Thanks also to Henry Iles for design, Bodhan Buciak for proofreading and Caroline Wilding for the index.

Above all we are grateful to our family, to whom we dedicate this book.

Alexandra Witze and Jeff Kanipe

Boulder, Colorado, January 2014.

Index

Page references for photographs, illustrations and maps are in *italics*

M

N

O

P